# The Poetic Structure of the World

Translated by Donald M. Leslie

# The Poetic Structure of the World

## Copernicus and Kepler

Fernand Hallyn

ZONE BOOKS · NEW YORK

1993

© 1990 Urzone, Inc.
ZONE BOOKS
611 Broadway, Suite 608
New York, NY 10012

Originally published in France as *La Structure poétique du
monde: Copernic, Kepler.* © 1987 by Éditions du Seuil.

Figures 3.3, 3.4, 5.1, 7.1 and 7.2 are from the University of
Ghent Library; Figure 3.5 is from the Bibliothèque Nationale
in Paris; Figure 5.2 is from the Vatican Museum; and Figure
3.6 is from the Biblioteca Apostolica Vaticana.

Printed in the United States of America.

Distributed by MIT Press,
Cambridge, Massachusetts and London, England

Library of Congress Cataloging in Publication Data
Hallyn, Fernand
    [Structure poétique du monde. English]
    The poetic structure of the world:
    Copernicus and Kepler/
  Fernand Hallyn
    p.  cm.
    Translation of: La Structure poétique du monde.
    ISBN 0-942299-60-4 (acid-free paper)
    ISBN 0-942299-61-2 (pbk.)
    1. Astronomy – Philosophy.  2. Science –
    Philosophy.  1. Title
QB14.5.H3513  1990
501 – DC20                               89-35295
                                         CIP

# Contents

*Introduction*   7

PART ONE   COPERNICUS OR THE RENAISSANCE
OF THE COSMOS

I   *Science and Irony*   35
II   *Science and Anagogy*   53
III   *"The Principal Consideration..."*   73
IV   *Synecdoches*   105
V   *Metaphor of the Center*   127

PART TWO   KEPLER OR THE MANNERIST COSMOS

VI   *From Copernicus to Kepler*   151
VII   *The Semiosis of the World*   163
VIII   *The Cosmographic Mystery*   183
IX   *Thinking the Ellipse*   203
X   *The Musical Metaphor*   231
XI   *The Lunar Dream*   253

*Conclusion*   281
*Notes*   287
*Index*   355

# Introduction

The existence of a poetics of science is undeniable.... Barring poetics from science is the same as barring use of the hypothesis.

— Melandri, *La Linea e il Circolo*

## *In Favor of a Poetics of the Hypothesis*

I intend to grapple with the "poetics" of Copernicus and Kepler. But what does this mean? What justification is there for applying the term "poetics" to the study of the scientific enterprise? What is there in that enterprise to which it can be applied? What relevance does the practice of poetics have for the epistemology and history of science?

### *A Placeholder*

I will not consider the logical form of theories, nor their truth value, but rather *the way in which hypotheses are established* as such. Of all scientific events, the "moment" that Charles Peirce called *abduction*, encompassing "all the operations by which theories and conceptions are given birth," poses the most problems for both the epistemologist and the historian of science.[1] It is true that factual connections sometimes furnish at least partial expla-

7

nations for the appearance of a new hypothesis: the observation of new phenomena, the development of new techniques, progress or a revolution in related disciplines.... But generally speaking, the establishment of a new hypothesis remains an enigmatic moment. Peirce himself spoke of an instinctual act, supposing a conformity between the laws of nature and the activity of the mind which develops under the influence of the same constraints it seeks to explain.[2]

The *epistemology* of logical positivism is not really concerned with how theories originate. Instead it focuses on the form of theories once they are established, for which it seeks to determine the criteria of "scientific acceptability" through an analysis of the operations of deduction and validation. The activity by which a theory is established is relegated to a prelogical stage, situated outside of the scientific enterprise properly speaking. Carl Hempel, for example, stresses that empirical "facts" are never logically relevant except in light of an already existing hypothesis; their relevance is never *a priori*. The way in which facts are selected and combined to establish a particular hypothesis depends on previous hypotheses and involves the play of the imagination in an indefinable act of creativity.[3] Karl Popper invokes literary or artistic *intuition*: hypotheses are "*free* creations of our own minds, the result of an almost poetic intuition, of an attempt to understand intuitively the laws of nature."[4] Often the formation of hypotheses is reduced to a psychological category such as association, intuition, or induction – in short, to the general conditioning that applies to any psychic or mental activity.

The *history of science* has long considered the succession of theories as a continuous phenomenon, composed of progress, correction and growth. Thus the principle of correspondence, so named by Niels Bohr, affirms that an earlier theory must at least be "contained" in the theory that replaces it. Opposing this thesis of continuity is a conception of history as composed of discontinuous phenomena. For Thomas Kuhn, continuity or progress occurs only during "normal" periods of activity, periods dominated by "paradigms" established and accepted by the scientific community.[5] The weight given to these research models and their implications is based not only on their logical consistency or their conformity to recorded phenomena, but also on institutional and sociological factors. Eventually, in response to a crisis brought on by anomalies that have become too numerous or phenomena that have become too substantial to ignore, another paradigm replaces the first, a new language takes over and speaks of another world. Between successive paradigms, there is no correspondence, but a logical break. Since the history of science is characterized by the "essential tension"[6] between tradition and innovation, we are concerned with two kinds of hypotheses. One enlarges the field in which an existing paradigm is applied, or contributes to its adaptation and local specificity. It does not touch what Joseph Sneed calls the "core of the theory" of the acquired paradigm.[7] On the contrary, a hypothesis of this type continues to affirm that core. In this case, the new hypothesis shows innovation within tradition. The steps leading to such a hypothesis presuppose a logical process within an accepted framework. But another kind of new hypothesis also exists, one that does not add to the "core of the theory" but replaces it. In this case according to Kuhn, "a relatively sudden and unstructured event," a "flash of intuition" takes place – the decisive turning point in a scientific revolution that cannot be rationally explained.[8]

In a manner that is sometimes more mitigated (e.g., Stephen Toulmin[9]) and sometimes more radical (e.g., Wladyslaw Krajewski[10]), some scholars have reintroduced the idea of a larger continuity between different moments in the history of science.

According to Dudley Shapere, in spite of abrupt breaks "there is often a chain of developments connecting two different sets of criteria, a chain through which a 'rational evolution' can be traced between the two."[11]

From an orientation that usually stresses the elements of at least a relative continuity, attempts have been made to recognize a *specific logic* in the formation of hypotheses. Norwood Russell Hanson had already defined a logic that establishes the plausibility of a hypothesis and a logic that justifies its acceptance.[12] Thereafter, investigators have chiefly sought to treat the specific logic of the hypothesis as a logic of questioning[13] and/or a logic of metaphor making.[14] But even if such a logic can accurately describe the sequence of questions and answers or the unfolding and implications of an analogy, the problem still remains of establishing exactly what determines the *choice* of the initial question or the first comparison, that is, the terms that are not part of the development but from which the development is launched. To be sure, it would be false to separate rigorously a problem and the existing state of an accepted theory. There can be a problem only to the extent that a theory *already* exists against which something constitutes an obstacle or *problema* (in the etymological sense). But that does not mean that the precise questions or the concrete analogies are dictated or constrained by theoretical givens that are already in place. The problem of determining the motivation of questions and metaphors cannot be fully resolved by such a logic. Within the context of a problem to be solved, choices are possible that do not uniquely depend on a theoretical state of affairs, even if one also considers "programs" or "traditions" of investigation.[15]

A method for studying the profound and subjective motivations that may lie at the base of these first questions and analogies – motivations existing outside of and in spite of purely logical

constraints — does exist: it is called *thematic* analysis. Conceived
by Bachelard as a "psychoanalysis of objective knowledge," this
thematic approach endeavors to study "how the fascination exerted by the object *distorts inductions*," the "extraneous flashes that
bedevil the proper illumination that the mind must build up in
any project of the discursive reason."[16] But if, for Bachelard, thematic attachments are, above all, obstacles to overcome — the supplementary *problemata* thrust between the mind and things — they
become for Gerald Holton the very basis for the activity of the scientific imagination, including its successes and positive results.[17]
The significance of this thematic illumination has been demonstrated by several detailed studies. But, since Holton's method
attempts to be "inductive and empirical,"[18] it proceeds by enumerating groups of particular themes, the organization of which
is not always clear. Holton recognizes the existence of "doublets,"
"triplets," and "multiplets" of themes, which leads one to suppose that some could be reduced to a single category or "arch-
theme." Some themes relate to the form of a theory (such as
"simplicity"), others to the elements that make up its content
(such as "atomism"); still others seem appropriate to metatheories
("complementarity"). If we are to get beyond empirical induction, it is undeniable that the application of Holton's thematics
must include a reductive effort directed toward providing structure. Holton himself recognizes the partial nature of any thematic
explanation; he combines it with sociological and institutional
analysis, and situates it in the specific context of the state of a
science at a given moment.

In fact, the scientist is not an "absolute" subject, the autono-
mous owner of specific capacities, whether logical, thematic, or
otherwise. The history of science studies how his actions are
linked to others in a specific community of scientists. But this
community in turn acts within a larger totality; its actions are

linked to other types of actions. We are correct, therefore, to turn to more general factors to explain the hypothesis.

This is the objective of the history of ideas, both the analytical variety (e.g., Lovejoy) as well as the synthetic (the *Geistesgeschichte* or, more recently, certain forms of the history of mentalities). The entire work of Alexandre Koyré, for example, is founded on "the conviction of the unity of human thought, particularly in its highest forms," linking scientific ideas to ideas that are "*transscientific*," philosophical, metaphysical, and religious.[19] Robert Lenoble speaks of "a *Weltanschauung* ... from which different representations of Nature emerge: the scientific, the aesthetic, and the ethical."[20] Faced with the hypothesis, there remains, however, the irreducible subject: "no one before Copernicus," writes Koyré, "had his genius."[21]

If the history of ideas explores a network of intersubjective relations, *ideological* analysis seeks the explanation for those relations in social structures. But we must admit that the sum total of representations, of value judgments and prescriptive judgments, all directly bound to that infrastructure and constituting an ideology,[22] cannot provide a mechanical and necessary explanation for specific constructions like science. That is not to say, of course, that science is free of sociological conditioning or ideological specification.[23] So when Jean-Pierre Vernant, for example, links the geometric astronomy of the Greeks with the *polis* of the sixth century, he in no way eliminates the need for supplementary mediations.[24]

Finally, Michel Foucault's *archaeology of knowledge* searches for determinations not at the level of a material exteriority, but through the analysis of discursive formations themselves: profound and specific constraints of an epoch ("epistemes") that lead to the production of homologous forms of discourse in seemingly unrelated domains.[25] But the object of this "archaeology" could hardly illuminate the concrete formation of a hypothesis. In Gérard Simon's terms, Foucault's method leads to the establishment of a *topography*, "the system put into place by the range of disciplines, the recognition of competencies, regularities of articulation, the intertwining of problems."[26] But the topography of places does not explain *what happens* there: "The individual who occupied that place was led to orient himself emotionally and intellectually, and the question is to determine how he did so. Therein lies an indispensable mediation, because we must pass from topography to action."[27] Here as elsewhere, when we look at the study of the formation of a hypothesis as a specific and global phenomenon, the need for a placeholder is *insistent*.

*The Meaning of a "Poetics"*
This overview of different approaches to abduction is certainly not complete. It suffices, however, to bring clearly into view the elements of a split: on the one hand, we see an accumulation of more or less direct and powerful determinations, and on the other, repeated confirmation of the irreducibility of those determinations. We do not intend to reinforce either of these tendencies. But, in the tension between determination and irreducibility, we must recognize the specific moment of the *choice* of a hypothesis (or of its elaboration) – neither entirely deducible nor entirely arbitrary.

A hypothesis, when it arrives on the scene and is not yet anything but a hypothesis, not yet true or false, arises from the realm of the possible. Its import is to indicate what may be in a more necessary, more systematic order. In this respect, Paul Ricoeur has drawn attention to the analogy between the role of the *mythos* in Aristotle's *Poetics* and the role of "heuristic fiction" in scientific research.[28] A new hypothesis, as long as it is not sufficiently validated and accepted, is formally similar to a *mythos*, to the intrigue or plot of a tragedy. For Aristotle, the plot is nothing other than "the arrangement of the incidents"[29] or "the organization of the events."[30] Now the goal of a hypothesis is precisely to organize "events" systematically. The systematizing hypothesis, moreover, is to naive perception what for Aristotle the work of the poet is to history: the hypothesis and the poem seek "events such as might occur and have the capability of occurring in accordance with the laws of probability or necessity,"[31] whereas naive perception and the historian record "what is" in the order of the accidental and the contingent. Anecdotal experience tells us that the sun revolves around the earth: this is what "takes place." The Copernican hypothesis, on the contrary, says that the earth revolves around the sun: this is what should take place in a more systematic, more probable, and more necessary order. Like the hypothesis, the *mythos* constructs reality. The hypothesis, when it is still only a possible existence, a hypothetical, systematic model, is this *mythos* which produces *mimesis* or "which composes and constructs that very thing it imitates!"[32]

This analogy with the Aristotelian *mythos* might appear to affirm our intention of calling the study of abduction a *poetics*. But of course we do not use the term *poetics* in the Aristotelian sense of a system of normative rules, but rather in the sense that one speaks about the poetics of Racine or Baudelaire, namely to designate a collection of choices made at different levels (style, composition, thematics...) by an author or a group. On the one hand, these choices lead to operations that inform the concrete work. On the other, they are loaded with meanings that more or less both determine and are determined by the artistic endeavor, for which the work is the result and sign. Ultimately, a study of poetics, in the sense understood here, comes down to what Umberto Eco calls "the plan for shaping and structuring the work."[33] It is the program for the execution of a work, informed by presuppositions and exigencies whose traces one can locate, on the one hand, in explicit declarations, and on the other, in the work itself, to the extent that its completed form, with respect to other works, gives witness to the intentions that presided over its production. A poetics must return to a way of dreaming works and the declarations that accompany them, of conceiving their possibility, and of working for their reality. Which imperatives led to this work? Which exclusions? How was it approached, and at what cost? These are some of the questions which I shall confront.

To speak of Copernicus' or Kepler's poetics in this sense will be justified only if it lends some specificity (over and above existing approaches) to the study of their projects, their choices and the processes involved. Instead of competing with the analyses of epistemologists or historians of science, or with the contributions of other disciplines, a poetics will take into account the illumination these analyses cast, and then examine the formation of a hypothesis as a global phenomenon that exceeds existing frameworks and is organized in its own manner — the result, certainly, of logical and other forms of determination and conditioning, but equally a unique configuration, an original synthesis that as a totality requires a study *sui generis* of the structure and interaction of the elements that frame and inform it.

Initially, I will differentiate three complementary tasks that a poetics must undertake: an inventory of topics, an intertextual insertion, and a tropological analysis. We will begin by presenting in broad terms the general characteristics of these tasks. Our aim here is not to elaborate a definitive and complete research program, but to provide the principal concepts and categories of which the concrete study of Copernicus and Kepler has revealed the operational validity.

## Commonplaces

According to the prevalent conception of rhetoric, common-places furnish a collection of topics for elaboration: "They are," writes Crevier, "general ideas applicable to a very large number of subjects, and offering openings for reasoning usefully about those subjects with respect to the end proposed by the Orator."[34] The very choice of which commonplaces to deal with dictates a thematic orientation. Why then can't we admit that science also is oriented by the kind of commonplaces that it proposes to deal with? Isn't an entire category of scientific activity – the category connected to abduction – determined by the selection and speci-fication of commonplaces?

The choice of certain commonplaces is inherent in Coperni-cus' and Kepler's choice of projects and procedures. Faced with a problem, they do not provide a direct solution, but trace lines of conduct, options, with a type of action in view. They restrict themselves to a small number of premises of a very general order, which point toward certain types of hypotheses and away from others. The choice of commonplaces initiates a research project, with imperatives that theory will be able to realize entirely or in part, just as the work of an artist or writer perfectly or imperfect-ly embodies the poetics that gives him direction. I will distinguish three sorts of commonplaces, related respectively to representa-tional schemas (*mimesis*), to the meaning of representations (*semiosis*), and to their truth value (*telos*).

### Mimesis: *Schemas of Representation*

In the Aristotelian sense of "the organization of events," *mimesis* undoubtedly constitutes the first goal of scientific activity. Now this arrangement presupposes a preference for certain schemas that direct the manner of *searching for* the order of things. As the most general presuppositional schemas, I shall retain the four

16

classes of "world hypotheses" described by Stephen Pepper, the validity of which has been demonstrated elsewhere: organicism, mechanism, contextualism, and formism.[35] These four visions comprise two antithetical pairs: the first pair is elaborated around the relationship between the whole and the part, the second around opposing conceptions of external causality. On occasion, I will use the results of more recent investigations to refine these definitions.

Organicism attributes specific qualities to the whole, qualities that cannot be reduced to the the simple sum of the parts. The "more" with which the whole is endowed may correspond to an immaterial principle, or it may be conceived monistically. In any case, it is a specific totality that conditions the mode of being of its parts according to the internal logic of its own coherence. The system predetermines its elements and their relations. In Pepper's terms, the organicist "is impressed with the manner in which observations at first apparently unconnected turn out to be closely related, and with the fact that as knowledge progresses it becomes more systematized."[36] Among different models of equal explanatory power, the organicist will tend to prefer the model that presents the highest degree of integration. He finally manages "to identify value with integration in all spheres."[37] For an organicist, the parts of a whole do not act; they have no relevant properties independent of one another. Hence it must be possible to study all the parts from the perspective of one and the same theory, and not by combining additive points of view or complementary theories.[38]

Mechanism, on the other hand, builds models out of individual parts. These parts maintain a degree of independence from the whole, which is never more than their result: "The higher level entities are described as combinations of lower level entities, but the entities on any one level are separate entities with their own locations in the space-time field and their own autonomy.... The universe is thus conceived as a huge aggregation or system of essentially separate individuals."[39] There are multiple entities which at each level of reality determine by their properties and interactions the appearance of reality. The whole, whether or not conceived as a system, is a product, not a controlling principle. Unlike organicism, mechanism allows the addition of theoretical points of view, applied to different parts and levels, if one can thereby infer the functioning of the whole.

The decision to accept or refuse explanations of phenomena based on final causes highlights the contrast between formistic and contextualistic criticism. The latter seeks a relational determination within a network of phenomena linked by spatiotemporal contiguity. It proceeds by establishing functional relations within an object or in its relationship with its environment. These relations are supposed to provide a sufficient context for explanation. Causality is conceived in terms of contact and transitive sequences in a system of determinations without a regulating sense of purpose. A contextualistic explanation does not rely on feedback mechanisms.

Formism, on the other hand, refers phenomena to a normative system. It calls upon a regulating principle that informs observed reality with a specific finality and activates feedback loops. It entails a process of selecting forms and ascribing value to certain relations, as Ernest Nagel writes, in terms of an orientation toward the acquisition or maintenance of a property or a state in spite of external or internal variations or perturbation.[40] Formism is linked to the notion of *program* or of "preordered signals that control a process (or behavior), leading it toward a given end."[41] In its *teleonomic* variety, formism consists of thinking in terms of a program that constitutes the final end and henceforth "gives direction to an otherwise aimless history."[42] The *teleological* variety of a formistic explanation invokes a program with a goal determined by its origin or its author. Today, formism is applied above all to living organisms and to human activity. But it supports equally well the ancient theory of "natural topics" [*des lieux naturels*]. And there can be a conflict, even today, between the state of a theory and poetic hope: Heisenberg held fast to the idea of an intention or plan, even in quantum physics, while also recognizing that "Such terms as 'intentions' or 'orders'... apply to human behavior and can at most serve as metaphors when applied to nature."[43]

Semiosis: *Schemas of Meaning*
The choice of a particular representational schema is often tied to presuppositions concerning the meaning science attributes to phenomena. Does the activity of signs — *semiosis* — that science must envisage play itself out in the system of directly or indirectly observable facts, or is it integrated into a higher order? Note that the question does not concern the meaning of the world in general, but only the meaning that is judged to be necessary and sufficient for the scientific enterprise.

Referring in this context to Léon Brunschvicg, Lenoble writes that "modern thought has created a type of *horizontal* explanation, in terms of equivalent effects and causes, all situated on the same level of empirical data, and this way of thinking has replaced the kind of *vertical* explanation that prevailed at an earlier time, and that linked visible effects to transcendent causes."[44]

A similar opposition appears in the semiotics of A. J. Greimas. "Consider eighteenth-century Europe," he writes, "which established the notion of natural sign and also thought of it as a reference to another natural sign: The sign *cloud* refers to the sign *rain*.... The interpretation of this type of civilization, which thinks of the natural world as the only level of reality," will only

seek horizontal explanations for phenomena. But other civiliza-
tions postulate a second level of reality and "interpret the sign
as a reference to this second-order reality" – in a vertical order.[45]

With Lenoble and Greimas, we can consider the seventeenth
and eighteenth centuries as the period of transition from the predominance of the vertical axis, linking several levels of reality,
to the predominance of the horizontal axis, reducing everything
to a single level. Both Copernicus and Kepler sought explanations
anchored in a vertical order: the world is the work of a divine
*poietes*, and their project implies that one can reach back through
the project to the Creator's *poetics*. What they aim to reveal
through their own poetics is thus truly, as my title says, the *poetic
structure of the world*.

A "vertical" meaning encompasses rather than eliminates the
task of also explaining so-called *horizontal* relations among phenomena. These are not taken, however, as pure systems of articulated facts, but rather as the signifying surface of a *code* that
correlates them to a transcendent signified. If thinking in terms
of a code wins out over thinking in terms of a system, the very
operation of *codification* can oscillate between *hypocodification* and
*hypercodification*.[46] Hypocodification assumes that in the absence
of more precise rules all or some of the facts can be ordered under
a common meaning as relevant elements of a potential code still
in the process of formation. Hypercodification, on the other
hand, adds supplementary types of correlation in the presence of
an already existing and more or less completely elaborated codification. Copernican "symmetry," which I will examine at greater
length later, offers a typical case of initial hypocodification, serv-
ing as a guide to the elaboration of an adequate world system,
the potential signifying surface of a code in the process of formation. On the other hand, Kepler's hypothesis in the *Mysterium
cosmographicum*, which brings into play the geometric code of the

curve and the straight line, is based on hypercodification, since it establishes *semiotic* relations that add to but do not replace "symmetry" in terms of vertical motivations of a different type.

Telos: *Valuation Schemas*

The *telos*, the final cause of a hypothesis, refers to the truth value of its representation of reality, to requirements concerning its referential scope, to its power to correspond to the real.

With "I assume there exists at least one object such that...," or "for all objects, I assume it is true that...," I can in fact appear just as much to pass judgment on reality as to propose a representation of reality whose eventually fictive nature I recognize, but which I decide to employ as if it corresponded to reality. I will not enter the properly epistemological debate over realism, conventionalism, and instrumentalism; my aim is not to advance that debate, but to situate two kinds of practical and antithetical undertakings anterior to established epistemological theories. These undertakings are based respectively on *irony* and *anagogy*.

Vico's *New Science* helps clarify the notion of *irony* in the sense that interests me here. For Vico, irony constitutes the last trope of "poetic logic," coming after metaphor, metonymy, and synecdoche. The order of presentation corresponds to Vico's hypothesis concerning the order of their historical emergence, according to which irony was the last to appear: "Irony certainly could not have begun until the period of reflection, because it is fashioned of falsehoods by dint of a reflection which wears the mask of truth."[47]

It is well known that Vico considered tropological language as the language of origins, the first language with which man discoursed on the world and phenomena. Now everything seems to indicate that irony, while an original figure, did not participate in the same origin as the other tropes. It requires a "reflected lie," consciousness of a distance, in contrast to the "simplicity" of the first ages.[48] Irony is a figure of origin not like the other tropes; it implies consciousness of the tropological nature of tropes. It does not first correspond to an operational schema intervening in the elaboration of "fables," but to a reflection on the elaborations that gave birth to the very notion of "fable." Irony is "metatropological" consciousness before it is tropological. It requires a critical stance toward discourse and its relationship with reality. It is only later that it becomes a trope like the others, thanks precisely to the consciousness of the *play* that is possible between words and things.

A figure for Vico of the origin of the age of reflection, metatropological irony continues to accompany this age throughout its history. Since Vico, it has continued under different names and changing conceptualizations to play an important role in the reflection on all forms of knowledge and assertions of knowledge.[49]

The irony that interests me here implies a self-reflexive judgment. As an evaluation of the act of proposing a hypothesis, it doubles that expository act with another act in which the subject assumes a certain distance from what he proposes. In this doubling, where the subject refuses to stand unequivocally by his own statement, irony shows both the refusal to give up expository discourse and the impossibility of assuming it totally. The interplay between statements about objects and reflection about these statements as objects is ambiguous. The subject is placed in a transcendent position with respect to his discourse, but only to deny the possibility of making himself the guarantor of transcendence.

In opposition to this irony, *anagogy* constitutes an enterprise for approaching knowledge of the ultimate principles of reality itself. With respect to the authors considered here and their belief in what I have called the "vertical" meaning of phenomena, the term *anagogy* usually takes on the religious orientation that also characterizes it in the medieval theory of exegesis. Medieval anagogy "leads to higher things" (*ad superiora ducens*)[50]; "supernatural" meaning is related "to the lofty things of eternal glory."[51] For Copernicus or Kepler, astronomy constitutes a way to God. They postulate the possibility of true knowledge of the celestial order and its motions as they depend on a higher and ultimate Principle that reveals or communicates itself in what it creates. The study of the world leads to the contemplation of God, as does the anagogical reading of the Bible. In his "Tentamen anagogicum," Leibniz enlarges the medieval meaning of the word "anagogy" by including exactly such a type of knowledge:

> Whatever leads to the supreme cause is called anagogical by philosophers as well as theologians.... The final analysis of the laws of nature leads us to the most sublime principles of order and perfection which indicate that the universe is the effect of an universal intelligent power.... The knowledge of nature... raises us to what is eternal.[52]

The opposition between irony and anagogy poses no difficulty in the context of the "vertical" meaning of phenomena. It is in that context that I will apply it here. But one can also apply it within the context of "horizontal" meaning. In his general theory of irony, Douglas Muecke differentiates between vertical irony and horizontal irony.[53] It is true that the notion of horizontal "anagogy" might encounter resistance. In certain cases, it is necessary to disengage the term from its traditional sphere of usage, to "secularize" it. In this spirit, Kenneth Burke speaks of a "socio-anagogic" dimension.[54] Such a yoking of words is justified if one accepts that the kernel of the term "anagogy" resides in the completely general idea of hierarchical directionality, of movement toward a hidden, enigmatic principle – a kernel that could be maintained in spite of variations in the referent to which the principle corresponds. "Anagogy" can therefore designate the movement toward any hidden principle, conceived as the ultimate mode of being.

## Cultural Insertion

The choice of commonplace topics is important, not only because it indicates at the outset a certain number of requirements that the enterprise must attempt to satisfy, but also because it constitutes a *differentiating* element in any given field: the adoption of this choice rather than some other makes it possible to account for fundamental dissimilarities. On the other hand, such a choice also constitutes an *integrative* element, making it possible to establish a link between a scientific enterprise and the whole culture of an epoch. Since it does not depend on established, well-defined theories, but informs and is presupposed by theory, it can also inform other practices in other fields.

A study of the poetics of a scientific enterprise leads therefore to establishing relationships with other cultural domains.

### Intertextuality

The history of science deals first with documents in their discursive dimension, that is, in terms of the complementary or competitive influence that they aspire to exercise in a limited disciplinary field: the space of positions and oppositions, continuities, conflicts, and breaks.

I mentioned above some broader points of view, most notably those pertaining to the history of ideas (Koyré, Lenoble) and the archaeology of knowledge (Foucault). Placing myself neither exclusively at the level of intersubjective influences, nor at the level of pure topographical reconstitution, my point of view will be *intertextual*, that is, it will entail both the study of the relationship of a work to particular texts, whatever discursive field they belong to, and the study of "the relationship between a text and the various languages or signifying practices of a culture."[55] This study is not limited to the field where an enterprise is commonly understood to operate; it will especially undertake to identify presuppositions that are *general* (those that dominate many or all the discursive fields of an epoch) and *heteronymous* (those borrowed from another discursive field to which they properly belong). And so the possible choices that I have enumerated can be presupposed and activated by different kinds of discourse. The existence of horizontal and/or vertical meanings, the preference shown for organicism or mechanism, the formistic or contextualistic option, ironic resignation or anagogical aim − all these can be found in philosophical or historical treatises, in literary fictions and artistic productions. These possible choices do not lead an abstract existence; they are conveyed by concrete texts. They are schemas that are integrated into the universe of a culture's codes and that are played out in the works, however diverse, which preach them theoretically or exemplify them practically. Some works and some fields can propose particularly striking and influential concretizations of such a choice. A scientific enterprise may also take up the topical presuppositions of works belonging to other domains. In the process of adapting them to its own purposes, it may even rewrite those works and inscribe traces of them in its own activity to such a degree that it is no exaggeration to say that those works become a precondition of its very possibility. The relation between the organicism and formism of Renaissance art and the Copernican enterprise is an example of such a case.

An intertextual study leads, therefore, to the insertion of various enterprises into a culture. It is in intertextuality that one can finally take hold of the connections, the inferences and interferences that establish the unity and diversity of a *culture*. My study of how a scientific enterprise belongs to the Renaissance will not attempt to subsume it in a static manner under a definition, but to describe how it is integrated, dynamically, in a network of presuppositional relations with other texts.[56]

Having recourse to the notion of intertextuality does not lead to what is sometimes called an "external" approach to science, in contrast to an "internal" study. The "formistic" conception of the circle, which we will find in Copernicus, is not "external" (philosophical, aesthetic) except for some modern historians of astronomy; it is not at all external for the author of *De revolutionibus*. Although they can be shared or borrowed (general or heteronymous), intertextual presuppositions are always internal to the enterprise they inspire.

### More than Imaginary Parallels

It is true that the intertext is an indefinite object. It is not only constituted by concrete and precise texts, but also by "systems... which can no longer be attached to their original texts"[57]: what is already written and already read has become anonymous, has lost its origin, and can be known only through a network of traces. However much an intertextual reading is interested in relations between single enterprises and texts, it is also interested in relations involving a presuppositional *reserve* that no longer belongs to anyone, but whose traces are visible in a series of given texts.

To designate intertextual relations that are not "causal," cases where the author of a text is not necessarily familiar with a given earlier text, or even anticipatory cases where one text seems to presuppose another, later text, Severo Sarduy has introduced the term *retombée*, which he provocatively defines as "timeless causality, non-contiguous isomorphism, consequence of something

26

that has not yet happened, resemblance to something that at the moment does not exist."[58]

Less paradoxically, it seems to me that *retombée* can be defined as the production of analogous effects from common presuppositions forming part of the anonymous intertext. If there is no causal linkage between the construction of churches with a central altar during the Renaissance and the central place of the sun in the Copernican universe, and if the second fact seems rather "to explain" the former – whereas it happened later – we escape the paradox if we recognize in both cases the same presuppositions concerning a vertical *semiosis*, the relationship between God and men. *Retombée* implies therefore a mediated relation: two non-contiguous positions, not influencing each other directly, but sustained by the same presupposition, two (or several) events "falling" from the same point. This explanation is similar to the willingness in divination to permit current facts to presage future phenomena or already resemble what does not yet exist: for Cornelius Agrippa, omens "predict coming things, not as causes, but as signs...which are related to them because *they arise from the same cause*."[59] By attaching parallel events to a common presupposition to which they refer, we demonstrate that the parallels are not purely imaginary, that they are not abstract and gratuitous correspondences attributable solely to the observer.

## Tropology

When we shift our attention from the very general choices mentioned above to their concrete effects in the cultural context of an epoch, what we encounter are actual tropological operations.

### A Restricted Tropology

The trope most often cited in the study of abduction is the metaphor, which is usually associated with the notion of *model*. Max

Black combines the terms in the title of an important work: *Models and Metaphors* (1962). Mary Hesse ends her book on *Models and Analogies in Science* (1966) with a chapter in which she replaces "model" with "metaphor": "the deductive model of explanation should be *modified* and *supplemented* by a view of theoretical explanation as metaphoric redescription of the domain of the explanandum."[60] Returning in 1977 to his earlier study, Black underlined even more emphatically "the tight connections between the notions of models and metaphors." A model is nothing other than the explanation of a metaphor: "Every 'implicative complex' supported by a metaphor's secondary subject...is a *model* of the ascriptions imputed to the primary subject: every metaphor is the tip of a submerged model."[61] Still more recently, Michel Meyer has affirmed that the "first level" of scientific activity, which culminates in the "problematic inference" of a hypothesis, possesses its own logic. This logic resides in the making of metaphors, and Meyer investigates this activity in the tradition of Hesse, among others.[62] In another recent study, Dedre Gentner juxtaposes texts by Galileo, T. S. Eliot, and Shakespeare, concluding that scientific analogy and literary metaphor are "more alike than different": the differences are of degree, orientation, and coherence, not composition.[63]

A metaphor does not always have the same status. With Michel De Coster, I shall differentiate a *discursive* status (valid in the case that aims to enlighten or convince), a *methodological* status (implying a heuristic function), and a theoretical status (linked to a vision of the world that poses *a priori* the existence of a real analogy).[64] It is clear that only the two latter types belong in a poetics of the hypothesis. A metaphor is discursive when it is applied to persuasion and exposition. The theoretical status applies especially to what Hans Blumenberg has called "absolute" metaphors, which are the "first elements of philosophical language, irreducible to the realm of logical terminology."[65]

When applying tropological schemes to abduction, there is a tendency not only to recognize or accentuate the role of only the metaphor (the preceding enumeration of references demonstrates this), but also to subsume other tropes within this category. Thus Meyer applies the term "metaphor" equally to metonymies and synecdoches, which appear as variants of one and the same figure. The statements "Richard is a lion" and "Thirty sails floated on the horizon" are treated in the same manner, and both are considered as abbreviated comparisons: "Richard fought as if he were a lion; thirty sails floated as though they were the boats themselves."[66] Such a "restricted" tropology risks telescoping different mechanisms that we have every reason to differentiate.[67] A diversified terminology permits us to achieve a diversified perception of facts. I will describe mathematical aspects of the Copernican proposition in terms of *metaphor*, its optical implications in terms of *metonymy*, and its physical innovations in terms of *synecdoche*. I will also have occasion to emphasize the *oxymoronic* aspects of Kepler's development.[68]

## Tropological Tropology?

If we attempt in the manner just described to avoid at a first level "metaphorocentrism," the question still arises as to whether we have eliminated it altogether: if metaphor isn't present at the surface, perhaps it is still present, but at a deeper level. What more is an interpretation in tropological terms than a vast metaphor? Isn't the reliance on tropological categories itself tropological? Two responses are possible, one weak, the other strong.

The "weak" response consists of being satisfied in affirming that the operations of abduction are *formally* similar to those implied by tropes, without prejudging their identity or the real relationship. Tropological categories are then descriptive instruments that are themselves metaphorical, at least provisionally. Reliance on these instruments has the advantage of exposing certain analogies, certain relations, the nature of which must still be studied in depth. The choice of instruments is based on criteria that are above all heuristic.

The "strong" response would appeal to the existence of a *deep* tropology at work in myths (Vico), language (Nietzsche), desire (Lacan) – an interior tropology that shapes the general nature of our representations of reality: "No doubt certain operations of displacement, of substitution, metaphoric or metonymic, have permanently marked the human *logos*, even when this *logos* has become a positive science."[69] Let me attempt to articulate more precisely this strong position, which is mine.

Tropes correspond to the formal procedural models that the subject can adopt during the conceptual transformation of a conventional representation. The theory of tropes is a theory of abandoning conventional semantics and of innovative representation. I will suppose that on that count the theory can be generalized. The literary trope appears at once as distance – with respect to the usual code – and as reduction of distance – with respect to the experience of things, which it aspires to represent more exactly.[70] Scientific innovation presents the same double aspect. The principal difference resides in the orientation of experience: in terms borrowed from Piaget, we will say that it is above all a matter of the "assimilation" of things to the subject in one case, and the subject's "accommodation" to reality in the other.[71] These two activities enter into variable relationships of predominance or actualization or reciprocal potentialization. From this point of view, the specific effect of a literary trope lies in its manner of posing an "assimilation": it is interpreted and evaluated not with respect to how it conforms to objective reality, but as a subject's subjective assimilation of reality; it tends to "subordinate"

– eventually through the mode of fiction – "the environment to the organism as it is."[72] A trope in a scientific hypothesis also assumes the fresh and different look of the subject. But through the trope, the subject aims to subjugate himself and others to the constraints that the object is supposed to impose on all. "Bad" scientific tropes, such as Bachelard has studied, are those where the preponderance of a deforming assimilation blocks that subjugation. Literary tropology differentiates between subjects and objects. It also assumes that a tension is *maintained* with other forms of expression: the very difference is a sign of subjectivity. Scientific tropology, on the other hand, aims at supplanting, without always succeeding, the language pattern that it opposes: ideally, the relationship is one of substitution, where the new hypothesis, duly confirmed and accepted, introduces a new *proper* language.[73] But this return to what is proper does not interest us here. That is precisely the moment when poetics gives way to epistemology and the history of science.

# PART ONE

# Copernicus or the
# Renaissance of the Cosmos

Correct proportions give rise to correct
form, not only in paintings, but in all higher things,
however they are produced.

– Dürer

CHAPTER ONE

# Science and Irony

The most profound poetic change from Ptolemy to Copernicus concerns astronomy's *telos*, its ironic or anagogic schemas for appraising value.

The history of irony as a companion to the history of science has not been written. It is beyond the scope of the current study even to introduce that history; we can only point out some landmarks leading to Copernicus, while showing that an ironic form of consciousness remained remarkably stable all the way from Ptolemy to the scholastics and even to some representatives of the Renaissance. This irony is "vertical." It measures human knowledge against an absolute knowledge, according to which it is worthless, and which it can never attain.[1] The following chapter will describe the profound transformation that Copernicus' conceptualization of the relationship between man and God represents.

### The World and the Cave

According to the *Timaeus*, God is "the maker and father of the world": "poietes kai pater tou pantos" (28c). The origin of the world assures its beauty and perfection. "Now it was not, nor can it ever be, permitted that the work of the supremely good

35

should be anything but that which is best."[2] But in the absence of its father, the world is open to interpretation, and thereby suffers the fate of writing and painting, as evoked in the *Phaedrus*:

> The fact is, Phaedrus, that writing involves a similar disadvantage to painting. The productions of painting look like living beings, but if you ask them a question they maintain a solemn silence. The same holds true of written words; you might suppose that they understand what they are saying, but if you ask them what they mean by anything, they simply return the same answer over and over again. Besides, once a thing is committed to writing it circulates equally among those who understand the subject and those who have no business with it; a writing cannot distinguish between suitable and unsuitable readers. And if it is ill-treated or unfairly abused it always needs its parent to come to its rescue; it is quite incapable of defending or helping itself.[3]

Like writing and painting, the world is on display; like them, it returns "the same answer over and over again." It "circulates equally among those who understand the subject and those who have no business with it": physicists, mathematicians, and others. It is the object of diverse interpretations; it is "incapable of defending or helping itself."

Orphaned, a text open to interpretation without the help of its parent: this situation had already led Ptolemy to adopt an attitude of epistemological resignation. According to the *Almagest*, the astronomer is, in the final analysis, as far from knowledge of the reality of celestial movement as the men in Plato's cave are from the world of ideas:

Let no one, seeing the difficulty of our devices, find trouble-some such hypotheses. For it is not proper to apply human things to divine things nor to get beliefs concerning such great things from such dissimilar examples. For what is more unlike than those which are always alike with respect to those which never are, and those which are impeded by anything with those which are not even impeded by themselves?[4]

Divine-human, high-low, stability-change, constraint-liberty: all of the oppositions that Ptolemy brings into play are archetypal renditions of vertical irony.[5]

For Ptolemy, the human being finds himself at the bottom of the vertical axis of the world, which is also an axis of knowledge. To the degree that he is aware of his impotence, the astronomer appears to himself as an ironic "victim," the objective victim of his condition at the bottom of the world. We can draw an analogy between this perception and the ironic mode in literary fiction. It is not going too far to say that in the book of the world and from the perspective of the gods, the astronomer resembles the kind of character whom Northrop Frye calls "inferior in power or intelligence to ourselves, so that we have the sense of looking down on a scene of bondage, frustration, or absurdity...."[6]

To be sure, Ptolemy also praises the astronomer, who opens up "the way to the theological."[7] But knowledge of the theological is essentially knowledge of privation and leads to an ironic attitude toward speech, as appears most notably in the dual conception of *simplicity*.

Ptolemy assigns astronomy the goal of describing celestial phenomena in a manner that does not necessarily take account of their real explanation: "But it is proper to try and fit as far as possible the simpler hypotheses to the movements in the heavens; and if this does not succeed, then any hypotheses possible."[8] The criterion of simplicity intervenes most notably when he describes the sun's motion. Ptolemy prefers the hypothesis of an eccentric circular orbit to that of an epicycle, because it "is simpler and completely effected by one and not two movements."[9] Simplicity of astronomical representation corresponds to economy of thought, and no more. It is essential not to confuse the astronomer's representation of motion with the true simplicity of celestial movement: "Or rather it is not proper to judge the simplicity of heavenly things by those which seem so with us, when here not even to all of us does the same thing seem likewise simple."[10] The simplicity of a representation of the heavens always falls short of the real simplicity of the heavens. If astronomy leads to a form of theological knowledge, it does so by negative implication, through an ironic "reading" that leads to a meaning that is the opposite of what is said. "Simplicity" slips away through the constructions that attempt to exhibit it.

The same ambiguity arises with respect to the notion of *truth*. In the first pages of the *Almagest*, Ptolemy extols the certitude of mathematical knowledge. But this certitude is completely within the intellect. If the astronomer chooses a certain combination of eccentric orbit and epicycle, such and such a motion – necessarily – results. But nothing assures us that the real motion of the planets takes place by means of such combinations. Aristotelian physics even renders them improbable. Truth in the "noetic world of theory"[11] may be a fiction in the external world.

The conception of astronomy that we encounter in Ptolemy found its classical expression in a text by Geminus:

For example, the astronomer asks why the sun, the moon, and the other wandering stars seem to move irregularly. Now whether one assumes that the circles described by the stars are eccentric or that each star is carried along by the revolu-

tion of an epicycle, on either supposition the apparent irregularity of their course is saved. The astronomer must therefore
maintain that the appearances may be produced by either of
these modes of being; consequently his practical study of the
movement of the stars will conform to the explanation he has
presupposed. This is the reason for Heraclides Ponticus' con-
tention that one can save the apparent irregularity of the
motion of the sun by assuming that the sun stays fixed and that
the earth moves in a certain way. The knowledge of what is
by nature at rest and what properties the things that move have
is quite beyond the purview of the astronomer. He posits,
hypothetically, that such and such bodies are immobile, certain others in motion, and then examines with what [addi-
tional] suppositions the celestial appearances agree.[12]

The ironist always says one thing *and* its opposite. Geminus does
not simply reject mathematical astronomy. While emphasizing
the uncertain character of astronomical representations, he does
not at all suggest that we should stop constructing such models.
At one and the same time, he affirms their value (for "practical
study") and their non-value (on the level of correspondences with
reality). Neither of these affirmations destroys the other. Irony
is the position that permits the union in one perspective of the
yes and the no.[13]

## A "Negative Mysticism"

### Oresme and the Earth's Rotation
The attitude that expects no more than utility from an astronomical representation persisted throughout the Middle Ages.[14]
Mathematical astronomy studied celestial phenomena in their
changing combinations, for which it furnished representations

that were always provisional (and required more and more *ad hoc* adjustments), with no pretence at referring to a deeper reality based on their source in a divine and necessary order. Especially after versions of Aristotle and Ptolemy became available in Latin translations, we again see fairly precise discussions of the question.[15] Thomas Aquinas affirmed, as had Averroës, that one cannot demand truth of an astronomer's constructions.[16]

Hans Blumenberg remarks that Oresme shows the "Nominalist's affinity for an inaccessibility of celestial processes."[17]

> T.: We must attempt from a far remove to find an answer to this problem.
> G.: [Aristotle] says *from a far remove* because the heavenly bodies are so far away from us.
> T.: They are not so far from us in the sense of spatial distance as they are in our limited perception of their sensible attributes.
> G.: Our only acquaintance with them is visual and from afar.[18]

This "inaccessibility" is illustrated concretely in the celebrated discussion of the earth's rotation. Let us recall at least the conclusion, which is extremely complex:

> However, everyone maintains, and I think myself, that the heavens do move and not the earth: For God hath established the world which shall not be moved, in spite of contrary reasons because they are clearly not conclusive persuasions. However, after considering all that has been said, one could then believe that the earth moves and not the heavens, for the opposite is not clearly evident. Nevertheless, at first sight, this seems as much against natural reason as, or more against natural reason than, all or many of the articles of our faith. What I have said by way of diversion or intellectual exercise can in

this manner serve as a valuable means of refuting and checking those who would like to impugn our faith by argument.[19]

How are we to understand this conclusion? Pierre Duhem, who brought the text to light, does not say; he concentrates instead on the discussion itself, where he sees a very clear and solid defense of the thesis that the earth moves.[20] Likewise for Albert Menut, the arguments presented in favor of terrestrial rotation are more convincing than the arguments on the other side; the conclusion would thus be only an apparent retraction, "prompted at least in part by a failure of moral courage."[21] But Edward Rosen believes that Oresme finally rejected terrestrial rotation for religious reasons:

The argument to which he attached the greatest importance was the quotation from Psalms 93:1. His principal purpose was to defend against rational attack his religious faith, which he recognized was entirely or mainly irrational. Thus he was essentially a theologian rather than a scientist.[22]

Perhaps such a divergence of interpretations was the very effect Oresme sought. His conclusion can seem like a thoroughly ironic text, multiplicitously ironic, designed to permit diametrically opposed readings.

Retrospectively, the entire examination of the hypothesis that the earth moves is characterized as *play* [*esbatement*]. The text has an exemplary value: rational arguments must not make us abandon our confidence in the earth's immobility, any more than other arguments should be allowed to "impugn" the certainty of faith. The entire lengthy scientific discussion illustrates an experience relating to faith. Such a moralizing ploy to redeem the discussion risks appearing tainted with irony, since the reader is left ignorant of its "true" status until the very end. Not only does Oresme say one thing when he means another; everything is written in such a manner as to lead one to believe that the only meaning is the scientific meaning. The procedure can be explained, to be sure: to just the degree that the intrinsic solidity of the rational arguments is worked out in detail, the unshakeability of faith takes on its full value.

But Oresme's irony does not stop there. The allusion to Psalm 93 could not have the weight of a peremptory argument that Rosen assigns it, since, in the body of the text, Oresme advances the possibility of a non-literal interpretation of similar Biblical passages:

> One could answer the...argument, which concerns the reference in Holy Scripture about the sun's turning, etc., by saying that this passage conforms to the customary usage of popular speech just as it does in many other places, for instance, in those where it is written that God repented, and He became Angry and became pacified, and other such expressions which are not to be taken literally.[23]

It is equally appropriate to point out the ambiguity in comparing the choice of a model for the world and an act of faith. On the surface, the text declares that we can agree with the thesis that the earth remains at rest, despite counterarguments, just as we accept articles of faith in spite of rational objections. Nonetheless, as soon as arguments in either sense are no longer "obvious,"[24] the internal and implicit logic of the comparison also permits us to relate other astronomical choices to an act of faith. The process does not stop there. While comparing the belief that the earth is stationary to an act of faith, the text allows us to believe that the stronger resemblance is between faith and the other choice: according to Oresme, the *other* thesis is "as much against natural reason as, or more against natural reason than, all or many of the articles of our faith." Apparently therefore, the comparison with faith justifies the thesis that the earth is stationary; but in reality, it also says that adherence to faith is "as much...or more against natural reason" than the belief that the earth moves. And so while seeming to do the contrary, the comparison weakens resistance to a choice that superficially is rejected.

Thus, in his last "play," in his ultimate irony, Oresme's conclusion retreats from what it claims to have accomplished. This evasive movement still does not illustrate, however, the epistemological resignation that lies at the heart of scholasticism.

### Law and Contingency

The practical difficulties involved in astronomical study do not in principle prevent concrete exploration, leading even to positive knowledge, limited in scope though it may be. The fundamental epistemological resignation of scholasticism rests on the conviction that it is impossible to discover the workings of God's *potentia absoluta*, his absolute power, which has mandated the final causes of Creation. But if science cannot hope to elaborate a deductive and necessary cosmology, it can accumulate descriptive knowledge and study the manner in which the Creator displays his *potentia ordinata* as efficient cause. It is important, therefore, to distinguish the study of things that are above us (*quae supra nos*), objects of a curiosity that is legitimate but cannot always be satisfied, and the necessarily vain desire to know the intentions of God, the One who is above us (*qui supra nos*).[25]

Duhem, Maier, Crombie, Randall, Butterfield, and others have considered the distinction between *potentia absoluta* and *potentia ordinata*, along with the renunciation of the search for final causes in favor of the study of efficient causes, as the basis for a breakthrough in the experimental method.[26] According to this view, the late Middle Ages was ahead of the Renaissance in the scientific domain. However we take these value judgments – the Renaissance and Humanism do not lack defenders[27] – it is appropriate to note that the position of a Buridan or an Oresme continued to be based on a metaphysical presupposition according to which the origin of things is lost in fathomless divine liberty, limited only by the principle of non-contradiction. Since God in his *potentia absoluta* is impenetrable – man can never discern Creation's final cause – the hypothesis that the earth moves is just as worthy of examination as any other hypothesis:

> Thus, it is clear from what we have said that it does not follow that, if God is, the heavens are; consequently, it does not follow that the heavens move. For, in truth, all these things depend freely upon the will of God without any necessity that He cause or produce such things or that He should cause or produce them eternally.... Moreover, it does not follow that, if the heavens exist, they must move; for, as stated, God moves them or makes them move quite voluntarily. He demonstrated this action at the time of Joshua when the sun stood still for an entire day....[28]

Buridan also emphasizes divine liberty when he envisages applying the new theory of impetus to celestial bodies,[29] thus furnishing the foundation of a physics compatible with the earth's movement:

> And just as someone could imagine that it is not necessary to postulate that angelic beings move the celestial bodies, since we have no evidence in Scripture that we must postulate such spiritual beings. For one could say that when God created the

44

tice is founded on the consciousness of the incommensurability between God in his *potentia absoluta* and man constrained to examine only the *potentia ordinata*, the general practice of scientific discourse necessarily takes on an ironic connotation. The irony lies in the ambivalence that surrounds any proposed theory from the perspective of the metascientific view that circumscribes the scientific domain. An explanation can assume the function of *law* in the Creation in which we live, but from the point of view of the *potentia absoluta*, it is only a *contingency*. The world and its laws are fundamentally contingent realities. God could have not created them, or he could have created them differently; he still could, at any moment, change the established order – impart to fire, for example, a cooling tendency. What we human beings call law is contingency from the Creator's point of view. Scientific discourse in general, just as any given proposition concerning an object that we only know from "afar," appears *sub specie ironiae*. Northrop Frye's description of ironic narrative also applies to this conception of scientific activity: "the wheel of time completely encloses the action, and there is no sense of an original contact with a relatively timeless world."[34] The irony is due precisely to the position of a "timeless world" that science cannot penetrate and where the apparent meaning of phenomena is transformed into a different and contrary meaning. If it is true that Buridan and others (though not all others: consider Nicolas d'Autrecourt) asserted the possibility of attaining probable knowledge of the efficient causes of phenomena, it is important not to lose sight of the fact that their attitude is less that of the conqueror (as will be the case with Galileo) than of the exile. It is not by chance that scientific nominalism flourished at the time of mystics such as Eckhart and Ruysbroek[35]: the exile of reason for the former is the flight of the soul for the latter. In its positivism, conscious that it represents a renunciation of an anagogical reading of the world, the science of the nominalists appears under that form of ironic consciousness that Lukács calls "negative mysticism."[36]

### Irony and Skepticism

#### Two Forms of Skepticism

Ironic consciousness did not disappear with the Renaissance. On the contrary, it became the object of renewed reflection under the impact of study of the ancient skeptics. It is important to distinguish two currents of skepticism. One is Ciceronian, academic, but also influenced by Saint Augustine's *Contra academicos*; the other is Pyrrhonian, strongly influenced by the work of Sextus Empiricus.[37] These two currents are differentiated by the type of irony they express.

Skeptics of the first kind see "vain curiosity" in any study of the heavens. Following especially the tradition of Saint Augustine, this theme appears most notably in Erasmus and Melanchthon.[38] Skepticism toward astronomy appears also among Aristotelians in the tradition of Thomas Aquinas (Gaetano, Pereira) or Averroës (Nifo), and Platonists like Pontano.[39] The preface that Snecanus added to Peurbach's *Theoricae novae planetarum* probably furnishes the most obvious example. There, we read, for example, "the astronomers are obliged to assign so monstrous a form to the celestial spheres, that if they speak truly, it would be impossible to think of anything more deformed than heaven."

Far from stimulating the desire to promote a more satisfactory theory, these findings only serve as evidence of the "blindness of the human mind." Through vertical irony, astronomy becomes a lesson in humility. To signify the fundamental infirmity of the human mind, God makes heaven appear monstrous and places its real beauty beyond man's grasp.[40]

47

Such skeptical speech is characterized by the fact that it takes itself seriously in its critique of *other* forms of speech. Its irony is not self-reflexive. That is where it differs from radical Pyrrhonian skepticism. Drawing on a passage by Sextus Empiricus, Montaigne describes the opposition between the two forms of skepticism as follows:

> Clitomachus, Carneades and the Academics despaired of their quest, and judged that truth could not be conceived by our powers. The conclusion of these men was man's weakness and ignorance. This school had the greatest following and the noblest adherents.
>
> Pyrrho and other Skeptics or Epechists – whose doctrines, many of the ancients maintained, were derived from Homer, the Seven Sages, Archilochus, and Euripides, and were held by Zeno, Democritus, Xenophanes – say that they are still in search of the truth. These men judge that those who think they have found it are infinitely mistaken; and that there is also an overbold vanity in that second class that assures us that human powers are not capable of attaining it. For this matter of establishing the measure of our power, of knowing and judging the difficulty of things, is a great and supreme knowledge, of which they doubt that man is capable.[41]

The skepticism of the academicians at least maintains *one* affirmation, namely that it is impossible to arrive at certainty. The more radical skepticism, which also leads to the most total form of irony, does not presume to settle the question of the value of human knowledge and is "still searching for the truth." This variety extends skepticism to its own discourse and ends up with the impossibility of supporting any affirmation whatsoever.[42] Such an attitude does not necessarily lead to the abandonment of all research; it can just as well involve a renewed effort, since no affirmation, no negation is certain.[43]

This is the kind of skepticism that would appear to characterize Osiander's intervention in the publication of *De revolutionibus*, whether he was sincerely defending a belief or only engaging in a prudent maneuver.

### Osiander's Paradox

The surviving fragments of a correspondence between Copernicus, Rheticus, and Osiander reveal that, from 1540 on, Osiander advocated adding an introduction to *De revolutionibus*. This introduction would *deceive*, *please*, and *divert* attention, thus performing a triple seduction. It would mute *De revolutionibus*'s claim to truth, so as to disarm in advance potential adversaries, who would reject *a priori* any theory implicitly calling both Aristotle and a literal reading of the Bible into question:

> The peripatetics and theologians will be readily placated if they hear that there can be different hypotheses for the same apparent motion; that the present hypotheses are brought forward, not because they are in reality true, but because they regulate the computation of the apparent and combined motion as conveniently as may be. (3CT, p. 23)

Once any basis for controversy concerning the truth of the theory had been removed, it could please everyone solely on the strength of its beauty: "In this way they will be diverted from stern defense and attracted by the charm of the inquiry..." (*ibid.*). The reader would be "charmed" into an attentive examination, which would finally force him to reject other theories and acknowledge that *De revolutionibus* has the very non-fictional character that it refuses to claim: "their antagonism will disappear, then they will seek the truth in vain by their own devices, and go over to the opinion of the author" (*ibid.*).

According to Kepler, Copernicus refused this proposition with "stoical firmness of mind" (3CT, p. 23). In any event, when *De revolutionibus* appeared in 1543, it contained an epistle "To the reader, concerning the hypotheses in this work," which presented the Copernican system as a fiction, not bound by the alternatives of truth and falsehood, to be judged solely in terms of its effectiveness: "For these hypotheses need not be true nor even probable. On the contrary, if they provide a calculus consistent with the observations, that alone is enough" (OR, p. xvi). Unsigned, this epistle was often taken for Copernicus' own introduction. Because of the accounts by Kepler and Praetorius, we know that it was written by Osiander.[44]

This epistle evoked violent reactions. From 1543 on, Tiedemann Giese, Copernicus' friend, protested against what he considered a "fraudulent" accusation, even "calumny."[45] Kepler was just as severe in his assessment (OO, vol. 1, p. 246). He refused to believe that Copernicus had any foreknowledge that the text was to be inserted. We need not exhaustively review the reactions. Let us note, however, that attempts to justify Osiander also exist. Thus, Arthur Koestler believes that Copernicus knew of the epistle and had probably agreed to its publication.[46] More nuanced defenses have been proposed, notably by Oberman and Wrightsman.[47]

It is doubtful, however, that all these positions have been based on an accurate understanding of Osiander's text. Is it certain that the *Ad lectorem* corresponds purely and simply to a seductive enterprise envisaged in 1540? Is it certain that its unequivocal purpose is to give *De revolutionibus* the appearance of presenting a purely instrumental theory? To decide, we must do more than just read the epistle. We must examine its relationship to Copernicus' preface. Sixteenth-century authors have accustomed us, in any case, to much greater subtlety.

If Osiander's goal really was to make the work he prefaced appear as discourse with no claim to truth, we must admit he did so in a very incomplete fashion. To accomplish this goal, it would not have been enough to *add* a letter *ad lectorem*. Just as important, he would have had to *suppress* in Copernicus' text any passages that express the conviction of speaking the truth.[48] Osiander's negations do not cancel out Copernicus' affirmations. Rather they add to them so that, as Praetorius noted, a "clear" contradiction exists between the *Ad lectorem* and Copernicus' own letter-preface.

Osiander, therefore, does not nullify the claim of *De revolutionibus* to speak the truth. He maintains those claims while he contradicts them. The "Osiander effect" is a paradox: by turns, the claim to truth is both denied and affirmed.

Such behavior was not unique in the sixteenth century. It is to be found prominently in the *Apology* Cornelius Agrippa wrote for his *On the Uncertainty and Vanity of the Arts and Sciences*. The *Apology* begins by denying that the work makes any claim to truth, presenting it instead as a *declamatio*, that is "a work on a conventional theme, executed as an exercise, removed from the rules for determining truth, and not requiring assent."[49]

The similarity to Osiander's disclaimer for *De revolutionibus* is striking. Agrippa took cover behind the conventions of a genre, just as Osiander invoked, on the behalf of *De revolutionibus*, a generic and conventional conception of mathematical astronomy. And if Osiander's text is followed by contradictory affirmations from Copernicus himself, Agrippa declared himself ready to uphold the validity of any of his positions that have come under attack:

> Even though it is false and libelous to present these articles as assertions, and even though I can take cover behind the term

*declamatio*...I will not refuse to reply seriously to the counts of the indictment, and I accept the peril of the charges that have been brought against my book. Therefore, I am going to refute point by point this series of libelous articles.[50]

No one has commented better on such a strategy than Tournon: "In sum, he [Agrippa] pleads both irresponsibility and that he is within his rights." Agrippa claimed both to "offer a truth and to reject it, to affirm simultaneously that the ideas advanced in his work are upheld by specious arguments and that they are correct. As a consequence of this restrained denial, the validity of his message remains in question, and becomes the principal issue."[51]

"Restrained denial": we could apply the same characterization to Osiander's preface. As Praetorius' commentary makes clear, the contradiction between the content of the *Ad lectorem* and Copernicus' own letter-preface was obvious for the attentive reader.[52] Such a reader found himself confronting the same paradox as Agrippa's. Faced with a work whose truth value was both denied and affirmed, he had no choice but to assume personal responsibility for the significance he granted to the Copernican system. Copernicus' own text, nonetheless, speaks otherwise.

# CHAPTER TWO

# Science and Anagogy

## *Propter Nos*

If astronomical monstrosities illustrated, for some skeptics, the fundamental weakness of man, these same monstrosities also gave rise to the demand to reform astronomy. Thus Fracastoro protested against the Ptolemaic representation of the heavens and accused those who defended it of impiety. His objective was a *renovatio* of astronomy, a return to a more acceptable representation that agreed with Aristotle's physics. He wanted to bring the astronomical representation more in line with reality, to free it from Ptolemy's "fictions" and artifice:

> They understood the divine bodies in an erroneous, and even, to a degree, impious fashion, giving them positions and forms which do not in the least suit heaven.... But, in truth, it is against such individuals...that all of Philosophy, and certainly Nature, and even the spheres themselves have loftily protested. In fact, to this day there is no man worthy of the name Philosopher who has dared to place monstrous spheres among those divine and very perfect bodies.... Thus, since in this domain no one has been worthy, assuredly to this day Astronomy has remained monstrous and, for the most part, imperfect.[1]

Along with the desire to rid astronomical representations of their monstrous character (OR, p. 4) and to reinstate their claim to truth, Copernicus also understood the need to restore science's anagogical thrust. He verified that existing theories do not permit us to proceed back from the handiwork of the world to its Creator:

> For a long time, then, I reflected on this confusion in the astronomical traditions concerning the derivation of the motions of the universe's spheres. I began to be annoyed that the movements of the world machine, created for our sake by the best and most systematic artisan of all, were not understood with greater clarity by the philosophers, who otherwise examined so precisely the most insignificant trifles of this world. (OR, p.4)

In this passage, God is designated as *opifex*: "ab optimo et regularissimo opifice." The term reappears many times in what follows: "opificem omnium" ["maker of everything" (OR, p. 7)], "a divina providentia opificis universorum" [by "the divine providence of the creator of all things" (OR, p. 18)], "divina haec Optimi opificis fabrica" ["the divine handiwork of the Most Excellent artisan" (OR, p. 22)].... Copernicus' insistent recourse to this name clearly places astronomy in a cosmological perspective. From the same perspective, Copernicus insisted on the etymological sense of the words *caelum* and *mundus*:

> What is indeed more beautiful than heaven, which of course contains all things of beauty? This is proclaimed by its very names, *caelum* and *mundus*, the latter denoting purity and ornament, the former a carving.[2] (OR, p. 4)

Such superlatives, applied to the Creator as much as to Creation recall Plato's argument from sufficient reason in the *Timaeus* (30a), according to which the best of beings necessarily produces the best of works.

If it is true that by praising both the author and his work Copernicus implicitly refers to the *Timaeus*, the principle of sufficient reason subjects Creation to necessity, rather than making it the product of God's unfathomable freedom. This presupposition places Copernicus in opposition to Oresme and Buridan. But he also opposed Ptolemy, since rather than multiplying antitheses between the divine and the human, the high and the low, and so on, he postulated that this world was created *for us*, and *for our use*: "mundus *propter nos* ab optime et regularissimo opifice conditus," he declares in the passage just cited. As the beneficiary for whom the world was made, man can attain *true* knowledge. In place of a universe whose beauty and rationality escape us, and which thereby calls us to humility, Copernicus substitutes a cosmos for which man is the final purpose and whose true plan he can reconstruct (OR, p. 5): "I have no doubt that acute and learned astronomers will agree with me if, as this discipline especially requires, they are willing to examine and study, not superficially but thoroughly, what I adduce in this volume in proof of these matters."

Copernicus conceptualized the relation between God and man as a profound affinity, not a radical break. The preamble to Book One of *De revolutionibus* is particularly instructive in this respect. Astronomy is a "divine rather than human" science, we read. In practice, Copernicus staked everything on man's divinity (OR, p. 7): "it is highly unlikely that anyone lacking the requisite knowledge of the sun, moon, and other heavenly bodies can become and be called godlike."

As the accompanying reference to Book Seven of the *Laws* makes clear, the theme that astronomy is a road to God goes back to Plato. It is true that the same kind of Platonic overtone was already present in Ptolemy, as we have seen. But the author of the *Almagest* counterbalanced praise of astronomy with consciousness of the fundamental uncertainty of all mathematical representations. More than Ptolemy (or even Plato), the praise of astronomy, where Copernicus largely paraphrases Ficino's Latin translation of the *Laws*,[3] recalls Ficino's *Theologica platonica*. Like Copernicus, Ficino celebrated man's ability to reconstitute the world; he even went so far as to make a quasi-identification between man and God:

> There is one point that we must note above all others: it is not the first person to happen along who can discover how or why a work, constructed artfully by a clever artisan, was created so and so, but only someone who possesses an equal talent.... And so, since man has seen the order in the heavens, its progressions and proportions or results, how could anyone deny that he possesses almost the same genius as the author of the heavens and that he could, in a manner of speaking, create the heavens, if he found the instruments and the celestial matter, since he creates them now, in another manner of speaking to be sure, but according to a similar plan.[4]

The contrast between Ficino's optimism and Ptolemaic humility is obvious. Ficino represents the reaction of a non-specialized philosopher to the humility of the astronomer-technician who is conscious of the imperfections of his constructions. But it is significant that Ficino's faith in man's powers is also found in Copernicus, who, as technician, was attempting precisely to reconstruct astronomy in a manner that would reconcile it with cosmology and signify man's dignity and responsibility as beneficiary of a world created *propter nos*.

To be sure, this theme was already present among the Stoics, some Church Fathers, and in the hermetic tradition.[5] But the Renaissance brought it to the foreground as part of the reaction against scholasticism, and Copernicus extended its implications for the scientific enterprise. In the Middle Ages, developments on the theme of Deus *opifex* or *artifex* were based on cosmological conventions or convictions, unconcerned with mathematical astronomy.[6] Copernicus was not content to admire an inaccessible wisdom "from afar"; he believed that science must permit man to penetrate the arcana of the divine plan and must be willing to submit to complete reform if necessary to achieve this anagogical goal.

The theme of the *propter nos*, planted by Copernicus among the premises of mathematical astronomy, began to spread before Ficino. Petrarch stressed it in *De sui ipsius et omnium ignorantia* [*On his own Ignorance and that of Others*].[7] We hardly need recall the anthropocentrism of Pico della Mirandola in this context, but we should emphasize that expressions like *propter nos*, *propter homines*, or *propter hominem* also correspond to a fundamental theme in Lorenzo Valla: "Who could be so blind, so lacking in lucidity, not to understand that this universe was created *for us*?... Everyone confesses that the universe was made by God: it was also made *for us*."[8] Charles Bovelles developed at length the theme of man's double nature: man is a microcosm not only through the analogical relationship between his body and the world, but also and especially because he is capable of interiorizing the world as mental image. Man is outside of the rest of Creation (*extra omnia factus*) as its mirror (*Homo Universi speculum*) or beneficiary. Man representing the world to himself is like the apex of a triangle or the center of a circle: everything that exists outside that reference point is like the base of the triangle or the circumference of the circle – made so that he can see himself therein.[9]

Bovelles' comparisons reveal a measure of indifference to the precise geometrical or physical centrality of man: apex (of a triangle) or center (of a circle), the distinction is unimportant; what counts is that man is the fixed point with respect to whom and for whom a presentation is arranged.[10] Henceforth, if man is the beneficiary of the world, his profound "centrality" remains, wherever he is physically located. He always constitutes the center of intellectual re-presentation, the being around whom the circle of interiorized objects is arranged in at least potentially the proper order. Copernicus' universe, while removing man from the geometrical center, remains from this perspective profoundly anthropocentric.

The affair is not without a certain irony, however. Is there not a divine irony in the decision to create the world for man's use, but to create it in such a manner that its true (heliocentric) structure is hidden behind a deceptive (geocentric) appearance? Why this dissociation between the form of the world and the picture that we have of it? The question is not broached. On the other hand, whereas Copernicus believed firmly in the truth of his system, he did not believe that the essential work of astronomy was complete. Rheticus writes:

> As for Copernicus, even though he feared the unjust judgment of others and criticized very harshly his own discoveries, nonetheless, under pressure from the very reverend Tiedemann Giese, he wrote his work and allowed it to be published. He did not do it so that others would be satisfied by his discoveries and teachings, but rather to inspire them to move forward with all their energy.[11]

An ironic distancing from his own certainty is present in Copernicus, but it is held in reserve, not a dominant feature. It is true, as *De revolutionibus* declares – in agreement moreover with the Bible (*Ecclesiastes* 3:11) – that the complete truth itself is only accessible "to the extent permitted to human reason by God" (OR, p. 3). But the movement toward this truth – anagogical movement toward final causes – belongs to man, inscribed at least as possibility and obligation in the *propter nos* of the Creation.

## Alter Ptolemaeus: *The Palimpsest*

Copernicus' contemporaries were practically unanimous in viewing him as astronomy's great *restorer*. Even those, like Brahe and Clavius, who did not accept his system, agreed on this point and called him a "new Ptolemy."[12]

Copernicus' work is strongly marked by his respect for antiquity. The *Letter against Werner* affirms the necessity of remaining faithful to the ancients' observations and vehemently condemns those who do not:

> we must follow in their footsteps and hold fast to their observations, bequeathed to us like an inheritance. And if anyone on the contrary thinks that the ancients are unworthy in this regard, surely the gates of this art are closed to him.[13] (3CT, p. 99)

Copernicus takes care, moreover, to emphasize that the very theory he is proposing is based on an ancient hypothesis concerning the nature of the universe:

> I undertook the task of rereading the works of all the philosophers which I could obtain to learn whether anyone had every proposed other motions of the universe's spheres than those expounded by the teachers of astronomy in the schools. And in fact first I found in Cicero that Nicetas supposed the earth to move. Later I also discovered in Plutarch that

certain others were of this opinion.... Therefore, having obtained the opportunity from these sources, I too began to consider the mobility of the earth. And even though the idea seemed absurd, nevertheless I knew that others before me had been granted the freedom to imagine any circles whatever for the purpose of explaining the heavenly phenomena. Hence I thought that I too would be readily permitted to ascertain whether explanations sounder than those of my predecessors could be found for the revolution of the celestial spheres on the assumption of some motion of the earth. [14] (OR, pp. 4–5)

The importance of this passage lies not only in the way it recalls certain precursors, but also in the weight it ascribes to a particular form of *renovatio* based on the liberty to think, which may in turn lead to *innovatio*. This liberty is all the more justified since the facts at the disposal of the sixteenth century are more numerous than those known to the ancients: "For the number of aids we have to assist our enterprise grows with the interval of time extending from the originators of this art to us" (OR, p. 8). Ptolemy had already affirmed the same idea. [15] It also appears in the epigraph to the *Narratio prima*, along with a quotation from the *Didaskalikos* that reminds us of the idea's ancient origins: "Free in his mind must he be who desires to have understanding." [16] The liberty to formulate new constructions is thus another principle that belongs to the heritage of antiquity.

The marriage of respect and liberty, of *renovatio* and *innovatio*, is clearly expressed in the hypertextual relationship that joins *De revolutionibus* to the *Almagest*: one is grafted on the other not as a commentary but as a transformation, implying both similarity and difference, or better, similarity within difference. [17]

The contrastive definition of hypertextuality (transformation vs. commentary) is important: the transformation establishes a relationship between "hypotext" and "hypertext" that is tied to the Renaissance practice of imitation[18] and is opposed to the practice of commentary predominant among the scholastics. Buridan and Oresme presented their observations as "paratext,"[19] clarifying, scrutinizing the original text, juxtaposing their text with it in an attempt to define the possibilities and/or norms of its reading. Is the agreement total, or not? necessary, or not? Their own discourse only develops its virtualities to the extent that the work being commented on permits, and it does so in a necessarily fragmented manner.[20] Hypertext, on the other hand, envisages hypotext as a total phenomenon, toward which it adopts a global attitude: a phenomenon not to read, but to rewrite. The relationship is no longer that of center and periphery, but a "palimpsestic" relationship of substitution.

Kepler stressed that the author of *De revolutionibus* interpreted the ancients rather than nature (GW, vol. 3, p. 141). He especially noted Copernicus' concern "not to disorient the diligent reader by straying too far from Ptolemy" (GW, vol. 1, p. 50). This preoccupation explains why the center of planetary motion does not coincide with the center of the universe: distances are not calculated to the true sun, but to the center of the earth's sphere (the median sun), such that this sphere corresponds exactly to the ancient sphere of the sun. As for the language and the order of exposition, Copernicus follows faithfully the model of the *Almagest*. He

> duplicates not only the mathematical machinery of Ptolemy but also the method and structure of his book. Chapter by chapter it has the same format and language, the same arrangement of subject matter with only the slightest changes as dictated by the change to heliocentricity.[21]

Such fidelity recalls what Thomas Greene called a "sacramental" practice of imitation, following as closely as possible the example of a revered model, as in a ritual performance whose success depends on the exactness with which the original actions, accomplished *in illo tempore*, are reproduced.[22]

But renewing an existing model also procures space for its transformation. At the level of content, *De revolutionibus* aims at nothing less than to defend a model of the universe that the *Almagest* had specifically rejected. To this "transmotivation"[23] of the content, we must add a remarkable difference in form. *De revolutionibus* is divided into six books, rather than the thirteen of the *Almagest. Six*: the first perfect number according to the Pythagoreans, the number of the days of Creation according to *Genesis*; the number of the planets in the new heliocentric system.... Should we not consider the possibility of a symbolic foundation, especially given that Rheticus praised the number *six* in his *Narratio prima*?[24] And add to the transmotivation of content the transmotivation of form? While the partitioning of the *Almagest* was motivated solely by clarity of exposition, the partitioning of *De revolutionibus* corresponds to the model under examination. To be sure, Copernicus himself never said a word on the subject, but even this silence might correspond to another form of *renovatio*, namely the return to Pythagorean discretion, whose exemplary value Copernicus recognized in the letter-preface he addressed to the Pope, as well as in the *Letter from Lysis*, which closes the First Book of *De revolutionibus* in the manuscript.[25]

### Astronomy and Philology

Obviously, the practice of paratextuality did not disappear during the Renaissance. But it did undergo a significant change. Instead of engaging the original text in an ongoing process of interpretation and discussion from the perspective of next in

a line of successors, the paratext began to engage that text as
something to be reconstituted in its historical distance from the
present, beyond the screen of intervening paratexts, which often
"have no dealings with true doctrine."[26] The ancient, original
text must be purged of the corruptions inflicted by time and by
semantic or other transformations that risk distorting the authen-
tic meaning. A science of the text must precede exegesis. But
this methodological priority collides with the dogmatic priority
given to faith by the scholastics. Concerning Erasmus, Latomus
wrote: "he seems to put science before faith...whereas I, on the
other hand, put faith before science."[27] Copernicus foresaw the
same objection.

Born of respect for the ancient authors, philological research,
implying as it does the critical examination of texts and reliance
on documents, also tends to historicize the image of antiquity.[28]
As Pocock writes, "Documents tend to secularize traditions; they
reduce them to a sequence of acts...taking place at distinguishable moments in distinguishable circumstances, exercising and
imposing distinguishable kinds and degrees of authority."[29] By posing the requirements of documentary and documented knowledge, philology was certainly one of the most potent factors
contributing to the addition of the historical dimension to mythical relations with antiquity.[30]

Philology's relation to astronomy is not limited to the fact that
the one furnished the other with better texts, while contributing on a large scale to historicizing the picture of antiquity.[31]
Through its demand for liberty, its return to facts, to origins, the
philological attitude of the humanists is analogous to Copernicus'
attitude. Just as Lorenzo Valla, Poliziano, or Erasmus wanted
to restore the true texts of the ancients and the Bible, so Copernicus' plan was to reconstitute the true texture of the world. His
enterprise was motivated by the presence of an uncertain text:

I have accordingly no desire to conceal from Your Holiness that I was impelled to consider a different system of deducing the motions of the universe's spheres for no other reason than the realization that astronomers do not agree among themselves in their investigations of this subject. (OR, p. 4)

At the beginning of his *Miscellaneorum Centuria Secunda* (unfinished work, not published until 1978), Poliziano compared his philological work to Asclepius' restoration of Hippolytus' mutilated body:

Book 2 of Cicero's *De natura deorum* was found in a no less mutilated state than Hippolytus mutilated by his maddened horses. But then, according to the legends, the illustrious Asclepius reassembled Hippolytus' torn limbs, put them back in place, restored them to life – Asclepius of whom it is said that he was thereafter struck by lightning because of the jealousy of the gods.
As for me, what jealousy, what lighting could frighten me into backing away from my attempt to restore to himself the father of the Roman language and philosophy...?[32]

Copernicus opens the *De revolutionibus* with the image of restoring a body, and with the same insistence on the sacred dignity of the task (OR, p. 4). Even if his principal point of reference is, as we will see, painting (depiction of a body), and on a more general level aesthetics (the opposition between what is monstrous and what is beautiful), his reliance on the image of a body whose integrity must be restored suggests the close relationship in the Renaissance intertext between philology and Copernican thinking about astronomy. The philologist, the painter, the astronomer are all, to a degree, Asclepius; and Hippolytus is by turns the text, human beauty, and the form of the cosmos.

Like Copernicus, the philologists insisted strongly on the necessity of reading facts and sources in terms of their specific codes. This is the source of Erasmus' affirmation of the rights of grammar in the face of theology:

> Neither do I think that Theology herself, the queen of all the sciences, will hold it beneath her dignity to be attended and waited upon by her handmaid, Grammar; which if it be inferior in rank to other sciences, certainly performs a duty which is as necessary as that of any.
>
> ... it is truly a new dignity for divines, if they are the only people who are privileged to speak incorrectly.[33]

*Mutatis mutandis*, we find in Copernicus an analogous argument in favor of astronomy. Like grammar, astronomy is doubtless "of a lower rank than some other sciences,"[34] but it constitutes "the summit of the liberal arts... supported by almost all the branches of mathematics" (OR, p. 7). It is not concerned with "insignificant trifles" (OR, p. 4), but with "the most beautiful objects, most deserving to be known" (OR, p. 7). Erasmus' plea in defense of philology applies *a fortiori* to astronomy, which also claims its rights in the face of a theology that thinks itself "too great to bend" to its instruction:

> Perhaps there will be babblers [*mataiologoi*] who claim to be judges of astronomy although completely ignorant of the subject and, badly distorting some passage of Scripture to their purpose, will dare to find fault with my undertaking and censure it. I disregard them even to the extent of despising their criticism as unfounded. For it is not unknown that Lactantius, otherwise an illustrious writer but hardly an astronomer,

speaks quite childishly about the earth's shape, when he mocks those who declared that the earth has the form of a globe. Hence scholars need not be surprised if any such persons will likewise ridicule me. (OR, p. 5)

If for Erasmus some theologians could be "barbarians" and "desecrators," Copernicus characterized the modern avatars of Lactantius as *mataiologoi*. Erasmus brought this term into fashion with a long commentary in his *Annotations* on the New Testament. It applies to scholastic theology, which was concerned only with *nugae* and *qaestiunculae* ("trifles" and "insignificant questions").[35] In the dedicatory epistle where he used the term, Copernicus also attacked the "philosophers" who are occupied with "the most insignificant trifles" (*res minutissimae*), and who neglect the study of important questions like the nature of the world (OR, p. 4). Whether directly influenced by Erasmus or not, Copernicus employs the very vocabulary used by the humanists when confronting authoritarian theology: "The humanists naturally applied that epithet to theologians, cavillers, and gossip-mongers."[36]

Of necessity, the theologian should submit to the results of astronomy as to those of philology. Philology, like astronomy for Copernicus, is theology's "servant": the world remains sacred; its beauty must lead man to admire "the Maker of everything, in whom are all happiness and every good" (OR, p. 7). But the servant has the right to guide her mistress "by the hand," and her mistress can only respond by adopting her language. "Astronomy is written for astronomers" (OR, p. 5). Only knowledge of mathematical language initiates an individual to the book of the world, just as knowledge of the language of the original text is, according to Erasmus, the necessary precondition for interpreting Scripture. His commentary on the adage *illotis pedibus* is one of the most vehement texts in this respect:

66

sive order, Reiss cites the telescope. Curiously however, explicitly specifying the means by which the conceptually usable character of the sign was brought into play, he does not mention pictorial perspective, which would seem to be an equally valid model.[40]

The perspectivist paradigm makes it possible to specify the scope of analytico-referential meaning. What concerns us here is not the motivation of signs within a *semiotic* system, where "each sign enters into a network of relations and oppositions with other signs that define and delimit it within a language."[41] The issue does not concern language as such; it is not to the question posed by the *Cratylus* nor is it even analogous. Depending on the observer's point of view, a square will appear as a square, a diamond, or something else. Confronting the same scene, I can adopt a frontal or oblique perspective, a perspective from above, and so on, and the shape of the signs that compose it will depend on my vantage point. The pictorial sign is thus unstable: neither pure self-identity, nor pure difference from other signs. It possesses no distinctive marks within a "language." The association between signifier and signified is variable, occurring on the level of *discourse*, of the production – specific and unique each time it happens – of what Benveniste calls the *semantic*.[42] Representation is conceptualized in its relation to the subject, which is its source.

As soon, however, as a relation of object to subject is established, a necessity is created. Conceiving a painting in terms of the relation between a point of view and the intersection of a pyramid guarantees that the elements composing the representation will be submitted to an analysis that obeys the requirements arising from the relation of the subject to the painting. This relation alone furnishes the law, the *certa ratio*[43] that henceforth exists between all the elements in the representation. The representation necessarily implies an immanent rationality based on the possibility of relating each element to the others and to the whole. The painting is a representation endowed with the force of a sanction, permitting one to "know" each element as a relational element in the whole that one "sees." Perspective, in sum, imposes an order through which the painter appropriates the world and offers this appropriation to others. This is what Alberti called "knowledge by comparison."[44]

To the degree, however, that the representation does not really give the exact equivalent of subjective perception[45] or of things themselves,[46] the knowledge contributed by perspective is of a special type. Perspective makes the representation and only the representation mathematical. The representation is the only thing it offers "knowledge" of. The perspective turns back on itself, on its own constructions. Perspectivist knowledge is knowledge of perspectivist phenomena. Through Alberti's "window," it is not the world that is on display, but something that is *already* a representation.[47] Nonetheless, the Renaissance consistently attributed to paintings a cognitive function with referential meaning and the ability to transcend subjective visual impressions. Perspectivism implies the intervention of the understanding in the act of seeing. Only the understanding, as Ghiberti wrote, can "see" the nature of things, "through discrimination, or recognition, or proof."[48] The boldness of the perspectivist paradigm consists precisely in presenting this controlled access to the world as the real possibility of a valid rationalism embracing human experience.[49]

With philology, on the other hand, reflection on linguistic signs turns aside from the problematic nature of language as a system of Cratylean correlations between words and things. Philology is not concerned with the origin of meaning, but with language in action, the world of discourse in a succession of contexts. The philologist recognizes the variability of the associations between *verba* and *things*, but he also emphasizes that such associations become necessary or impossible in a given context. The philological enterprise attempts to construct a perspectivist representation of the past, to "know by comparing" in Alberti's sense – that is, to confine itself to the relations within a specific period in order to reconstitute a representation of the past where each element receives its meaning from its relation to other elements and to the totality of discourse as historically sanctioned.[50]

Even if linguistic meaning is conceived in a perspectivist manner, words can still lead to certain knowledge of the world and can still establish the truth of that knowledge. Man becomes aware of the world only through signs, but this world was created with man *in mind*. At the very least, the sufficiency of signs for the things they represent is possible, variable, perfectible. But there is no point in seeking this sufficiency by piling commentary on commentary, exegesis on exegesis, as if signs were endowed with a life of their own which we must support, develop, and sometimes correct or bring into line. To the contrary, we must define this sufficiency by recognizing the identity, the totality, the rationality of different historical situations. Moreover, we must reinvent this sufficiency from the new perspective of each new subject, who is a new beginning.

Valla declared that notions of truth and falsity must always be brought back to the "mind of the person speaking."[51] This is not the relativism of a skeptic. Instead it is a positive assertion that meaning is the ternary relation between sign, thing, and subject; and Valla went so far as to say that for man meaning and knowledge are one: "There is no difference between saying what wood is...and what 'wood' means."[52] Truth is produced by representation, which is not the passive reception of signs, but the production of speech. In this context, philological thought calls upon and provides profound justification for rhetoric or the *art* of speaking. Valla praised, for example, abundant, copious speech, but he does not mean by that the frivolous amusement of the idle talker. He attacks *loquacitas*, or the gratuitous multiplication of words and figures, which he distinguishes from *copia*, or the diversification of speech that produces a richer representation of the world.[53] Human speech produces the human world.[54]

Vivès, who argued both for and against Valla's new ideas,[55] and was urgently concerned with the truth value of the representations attached to words, made an explicit analogy with the constraint perspective places on the act of looking:

> We conceive of things through the portals of the senses, and, enclosed within the body, we have no other means to do so. We are like those who are in a room with only one opening through which we can see or light can come in....[56]

All knowledge, all judgments concern representation, but not necessarily what is represented:

> consequently, when we say that things are or are not, are this or that, are such and so, we base our statement on the judgment of our mind, not on the things themselves. Our mind is the measure, not things....[57]

Yet Vivès did not reject the possibility of positive knowledge of things. The distinction he drew between different degrees of knowledge (by the senses, by the imagination, and by the mind) enabled him to introduce degrees of certainty and to deny that when man has progressed through all of these, his judgment "deforms the truth of things."[58] Ghiberti, in a passage already cited, conceived the truth value of pictorial perspective in the same manner. It is enough to locate the criterion of truth in the rational and rationalizing operations of the mind: "modus cognitionis...nostrarum est mentium" [the measure of knowledge

71

is (the measure) of our intellects][59] — which obviously presupposes that the world conforms to the requirements of human rationality, or that it was created, as Valla and Copernicus said, *propter nos.*

And so the same idea-force appeared in the epistemological development of philosophy and in the theory of pictorial perspective. For both Valla and Vivès, perspectivism did not lead to a special appreciation of mathematics, the result no doubt of the specific object of their reflection: natural language. Copernicus placed himself in the same tradition, but with an emphasis on the local importance of mathematical language.

The *De revolutionibus* also presupposed the perspectivist dimension of signs. In Kant's celebrated terms, Copernicus' greatness consists in having "dared, in a manner contradictory of the senses, but yet true, to seek the observed movement, not in the heavenly bodies, but in the spectator."[60] Thus Copernicus considered man as the "source" of the apparent motion of the planets and the sun. Observed immobility and motion are not necessarily qualities of things seen, but must be approached within an investigation that includes the subject as the origin of possible error and truth. But if the subject has access to things only through representations in which he himself is implicated, how is he to distinguish between true and false representations? Two presuppositions make such discrimination possible. One, as we have already seen, consists in granting that the world is made for man, and therefore he is capable of knowing it, of discovering the point of view from which Creation is organized. The other, which we have still to examine, endows the form of the world with an internal necessity whose effects man must strive to find through the act of representing it. As we shall see, this process is based on what Alberti called "knowing by comparing."

72

# "The Principal Consideration..."

In the prefatory epistle of *De revolutionibus*, dedicated to Pope Paul III, Copernicus made it clear that his principal charge against the astronomical theories of the day was of a cosmological nature, concerning the harmony, structure, and beauty of the universe:

> Nor could they elicit or deduce...the principal consideration, that is, the structure of the universe and the true symmetry of its parts. On the contrary, their experience was just like someone taking from various places hands, feet, a head, and other pieces, very well depicted, it may be, but not for the representation of a single person; since these fragments would not belong to one another at all, a monster rather than a man would be put together from them. (OR, p. 4)

Copernicus draws attention to this passage by stressing that his critique of geocentric conceptions is directed toward the "principal consideration." In Chapter 10 of Book 1, he repeats that the principle governing relations between the parts of the universe is the "first principle" (OR, p. 21). Certainly historians of astronomy have always recognized the importance of these lines. The term *symmetry* especially catches their attention. According to Gérard Simon, how one interprets this word "directs in large part the analysis that one can make of the mind that presided over the composition of *De revolutionibus*."[1] Rejecting an interpretation based on the modern geometrical sense, he recalls the etymological meaning, the fusion of *syn* and *metria* signifying "the act of measuring together" or the "common measure." From that perspective, Copernicus' preoccupation with *symmetry* is related to his speculations on cosmic *harmony*. This etymological reminder is opportune. From the outset, it points to the deeper motivation underlying the Copernican enterprise, a motivation not limited to the history of astronomy, but bound up with general aesthetic tendencies during the Renaissance. Fundamentally, the comparison between the universe and the human body asserts that all reality is subject to an organistic principle, and this principle is mathematical. In the analysis that follows, Copernicus is magnificently blind to qualitative relations between elements (which he nonetheless enumerates: hands, feet, head...). Far from drifting into an "elementaristic microcosmism,"[2] or seeking correspondences between the parts of the universe and parts of the body, as did many of his contemporaries,[3] Copernicus cares only about the quantitative relations between elements. *Similitudo* is thrust aside, and *proportio* or *analogia* is reduced to its pure mathematical form. The preeminent model for such a priority can be found in Renaissance art, which conceived of an idealizing *mimesis*, or elucidation of the underlying principle that presided over Creation, as the discovery of a coherent system of mathematical relations.

## From "Monstrosity" to "Symmetry"

*Mechanism: The Geocentric Monster*
In the *De caelo*, Aristotle wrote: "The movements of the several stars depend, as regards the varieties of speed that they exhibit,

there to prevent, in fact, putting Saturn beneath the Sun...?" As a mathematical object, the geocentric universe has no depth. Astronomy could not determine the length of the radius joining a given planet to the earth; all it could do was specify the angular position of the planet against the celestial canopy.[6] The order of the planets did not enter into the mathematical investigation.

On the other hand, it is well known that in a geocentric model planetary motion is not a continuous or smooth progression, but comprises reversals and pauses. Hence the celebrated comparison to a dance.[7] What has been praised as capricious but rhythmic motion, however, really contributes, from a more rigorous point of view, to the disorder of the universe. To explain those apparent irregularities, Greek astronomers such as Eudoxus and Callippus had constructed a system of homocentric spheres that Aristotle adopted and modified.[8] Since the earth was at the center of these spheres, the distance of any planet from the earth should have remained constant. But the planets appear brighter (and therefore seem closer) during their backward motion. The inability to account for these "phases" led to the elaboration of other theories, associated especially with the names of Apollonius and Hipparchus. Ptolemy further elaborated their system of eccentric circles and epicycles, and added the equant.[9]

Ptolemaic astronomy accounted for the alternation of forward and backward motion by *epicycles*, or circles whose center rotates on the circumference of another circle, the *deferent*, which in turn rotates against the zodiac. The deferent could be *eccentric*, and in the case of Mercury and the moon, the center of the deferent even *moves*.[10] But if this elaborate system made it possible to maintain the circularity of motion, it still raised questions about the rationality of the universe — not only about the different modes in which the deferents were constructed, but also about the determination, dimensions, and number of epicycles. It was precisely the interpretation that Copernicus could give to complex and hitherto unexplained relations between deferents and epicycles that permitted him to speak about his discovery of a "marvelous symmetry" in the universe (OR, p. 22).

According to the Ptolemaic theory, epicycles followed two different schemes, depending on whether the planet was slower and more distant than the sun or whether it accompanied the sun in its orbit around the zodiac. For the "outer" planets, the center of the epicycle and the position of the planet were always on a straight line parallel to the axis joining the earth and sun. For the "inner" planets, the center of the epicycle was always on that axis (cf. Figure 3.1). The structural difference was tied to the fact that the angular distance between the inner planets and the sun never exceeds + or − 45° for Venus and + or − 24° for Mercury (both remaining always in close proximity to the sun), whereas the angular distance between the outer planets and the sun can reach 180° (they are in conjunction with the sun at the apogee, in opposition at the perigee). At first consideration, the sun seems to divide the sky into two distinct parts, but, as Copernicus noted, the distinction is blurred by the moon's behavior. Closer to us than the sun, it nonetheless moves away from the sun as do the outer planets: "Ptolemy argues that the sun must move in the middle between the planets which show every elongation from it and those which do not. This argument carries no conviction because its error is revealed by the fact that the moon too shows every elongation from the sun" (OR, p. 19).

In the geocentric model, the dimensions of the epicycles never seem to obey any concern for proportion. In Copernicus' terms (OR, p. 22), "forward and backward arcs appear greater in Jupiter than in Saturn and smaller than in Mars, and on the other hand greater in Venus than in Mercury." Venus, with an epicycle amounting to two-thirds of the deferent, provided another flagrant case

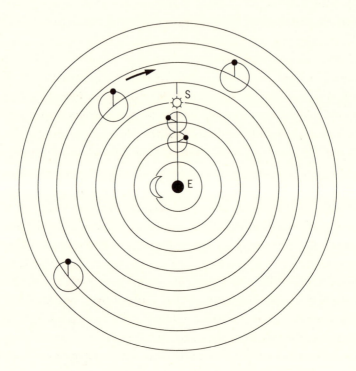

FIGURE 3.1. Construction of epicycles. For the planets between the Earth (E) and the Sun (S), the center of the epicycle is always on ES. For the outer planets, the vector of the planet as it moves on its epicycle is parallel to ES. In the first case, the deferent is a function of the relation between the earth and the sun, in the second case, motion on the epicycle is a function of that relation.

of nonconformity. Fracastoro remarked, "If by chance such an epicycle existed, its circumference would have to reach practically to the center of the sphere...."[11] Rheticus expressed his scorn this time with an ironic paralipsis: "I refrain from mentioning here the vast commotion which those who defame this most beautiful and most delightful part of philosophy have stirred up on account of the great size of the epicycle of Venus..."(3CT, p. 146). In theory, Venus should have presented very striking variations in size and brightness, which is not at all the case.[12] The moon was also a problem: the size of its epicycle was such that during the first and last quarters it should have been noticeably closer than during the new or full moon, which is contrary to observations of the angular dimensions of the lunar disk.[13] The number of epicycles per planet also seemed arbitrary and disproportionate. To account for observed phenomena, it was necessary to construct not only epicycles on eccentric circles, but also, in some cases, small epicycles on large, and so on.[14] This practice applied especially to Venus and Mercury, and Fracastoro criticized it severely.[15]

It seemed that Ptolemy could develop a suitable description for the motions of each planet, but he could not elaborate a *system* that explained the totality of phenomena in a coherent manner. The parts of the universe were passed in review, one by one, as elements in juxtaposition. Circular patterns were assembled to account for particular motions. The units constituting the path of each planet were described in terms of lower units, such as deferents and epicycles, without being subject to a totality provided with its own laws. Attention was given to relations between the parts and to analysis of those parts considered as individual configurations. The vision was *mechanistic*, hence vulnerable to Copernicus' organistic critique: parts, but not a body; a monstrous assemblage, as Galileo was to repeat: "But he [Copernicus] adds that, in going about putting together all the structures of the particular orbits, there resulted thence a monster and a chimera, composed of members most disproportionate to [one] another, and altogether incompatible."[16]

### The Metonomy of Perception

Although Copernicus also relied on a large number of epicycles, although Venus' epicycle remained very large, and although the systematizing was still whimsical in part,[17] he nonetheless established a "harmonious linkage between the motion of the spheres and their size, such as can be found in no other way" (OR, p. 22).

The establishment of this "harmonious linkage" (or "symmetry") required the reinterpretation of some empirical relations:

> Why should we not admit, with regard to the daily rotation, that the appearance is in the heavens and the reality in the earth? This situation closely resembles what Vergil's Aeneas says:
>
>> Forth from the harbor we sail, and the land and the cities slip backward.
>
>> For when a ship is floating calmly along, the sailors see its motion mirrored in everything outside, while on the other hand they suppose that they are stationary, together with everything on board. In the same way, the motion of the earth can unquestionably produce the impression that the entire universe is rotating. (OR, p. 16)

These lines pose the problem of the relation between perception and reality, illustrating the point with a literary example borrowed from the *Aeneid* (3.72). When Aeneas declares that "the land and the cities slip backward," his language is tropological;

he substitutes an effect for a cause. The "slipping backward" of the cities is a figure signifying the departure of the ships. Copernicus suggests that we may be misled by a similar "figure" when we look at the sky. In fact, any perception is the equivalent of a metonomy: an empirical relation or an effect (the content of the perception) always appears in the place of its cause (the state of affairs that provokes the perception). Whereas Vergil accounts for this metonomy by asserting the priority of the subject, who assimilates reality, the task of astronomy is to substitute, by an inverse metonymic operation, the cause for the effect. The perception of celestial motion is an effect, and we cannot decide immediately what cause it is substituting for. Based on this possibility for tropological deception, Copernicus proposed his redescription of the empirical relation between a terrestrial spectator and what he sees.

Copernicus' interpretation consists of considering the motion attributed to the sun in the geocentric model (cf. Figure 3.2) as a purely optical consequence of the earth's motion. In Kepler's summary:

> The first motion is that of the sphere, or circle, which in one year carries the Earth as a celestial body around the Sun.... The motion of this circle, or of the Earth, enables us to make an economy of three eccentrics in the usual hypotheses, namely those of the Sun, Venus and Mercury. Because the Earth revolves round these three planets, the inhabitants of the Earth are led to believe that these three [planets] revolve around a stationary Earth.... Furthermore, this circle [of the Earth] being admitted, the three large epicycles of Saturn, Jupiter and Mars disappear also.... [T]he Earth, viewed from Saturn (which is almost motionless, because it has the slowest motion), moves on its circle, first approaching and then reced-

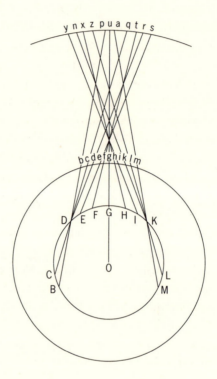

FIGURE 3.2. The Copernican explanation of retrograde motion as illustrated in Galileo's *Dialogue on the Great World Systems* (third day). The diagram describes an outer planet (Jupiter). While the earth moves from *B* to *E*, Jupiter advances from *b* to *e*, which is observed as a continuous motion in the zodiac from *p* to *s*. But when the two planets are respectively at *F* and *f*, Jupiter appears to have moved back in the zodiac toward *t*. The apparently retrograde motion continues until the planets are at *H* and *h* and Jupiter appears to have moved back to *x*. Reaching *i* (when the earth is at *I*), Jupiter appears stationary in the space *xy* and then seems to advance again in the same direction as the earth.

ing from Saturn; but the Earth's inhabitants believe that it is Saturn which moves on an epicycle with a backward and forward motion, whilst they themselves [the Earth] are motionless at the centre of its circle.[18] (GW, vol. 1, pp. 17-18)

*Organicism: The "Symmetrical" Body*
The Copernican interpretation revealed the necessity behind the striking numerical pattern that appears in the following table, a pattern that was gratuitous from the perspective of the geocentric model:

| Planet | Revolutions on the epicycle | Revolutions on the zodiac | Solar years |
|--------|------------------|-----------------|------------|
| Saturn | 57 | 2 | 59 |
| Jupiter | 65 | 6 | 71 |
| Mars | 37 | 42 | 79 |
| Venus | 5 | 8 | 8 |
| Mercury | 145 | 46 | 46 |

TABLE 3.1. Ptolemy's figures in *Almagest* 9.3.

Let us examine first the case of an inner planet. Venus seems to complete its epicycle 5 times in 8 years, which means, in terms of the heliocentric model, that during their common revolutions around the sun, Venus "joins" the earth 5 times in 8 years. Since Venus moves more rapidly than the earth,[19] the result is that Venus makes a total of 5 + 8 or 13 revolutions around the sun during 8 years. One revolution of Venus takes therefore $8/13$ of an earth revolution, or about 7½ months. Likewise, Mercury revolves 145 + 46 times around the sun in 46 years; one revolution lasts about 88 days. The agreement in the numbers in the last two columns of Table 3.1 (revolutions around the zodiac and solar years) is due to the fact that the deferent of a inner planet is nothing other than a transcription of the optical effect due to the earth's motion.

For the outer planets, since epicycles and not deferents correspond to the illusion produced by the earth's motion, the revolutions around the zodiac correspond to the number of revolutions around the sun: Saturn completes 2 revolutions in 59 years, Jupiter 6 in 71 years, Mars 42 in 79 years. Saturn requires almost 30 years to make one revolution, Jupiter 12 years, and Mars about 1½ years.

For the outer planets, Ptolemy had verified that the sum of the numbers in the first two columns in Table 3.1 equals the number in the third column.[20] Since, according to the heliocentric model, the epicycle of an outer planet is an optical effect produced by the earth's motion, the fact that Mars completes 37 revolutions on its epicycle while making 42 revolutions around the sun means that the earth has "joined" Mars 37 times. But to do so, the earth has had to complete 37 more revolutions around the sun than Mars for a total of 42 + 37 or 79 revolutions. The hitherto unexplained equality between the number of solar years and the sum of the revolutions on the epicycle and around the zodiac is thus a necessary equality.

On the other hand, nothing in the Ptolemaic system allows us to establish a necessary relationship between the dimensions of the deferents and the epicycles of the different planets (cf. Table 3.2). With Copernicus, the epicycles of some planets and the deferents of others are accounted for as appearances based on the earth's motion; this fact introduces a *general commensurability*. For each of the inner planets, the radius of the epicycle and the radius of the deferent are equivalent to the average distances from the planet to the sun and the earth to the sun. For the outer planets, the respective significance of the two radii is inverted: the radius of the epicycle corresponds to the average distance from the earth to the sun, whereas the average distance from the planet to the sun corresponds to the radius of the deferent. The ratio of average distances to the sun for Venus and the earth is therefore 43½ to 60, and for Mars and the earth is 60 to 39½. If we express the distance from the earth to the sun as 1 unit, the distance of the other planets from the sun is, in descending order, 9, 5, 1½, ¾, ⅓.

| Planet | Radius of the epicycle | Radius of the deferent |
| --- | --- | --- |
| Saturn | 6½ | 60 |
| Jupiter | 11½ | 60 |
| Mars | 39½ | 60 |
| Venus | 43⅙ | 60 |
| Mercury | 22½ | 60 |

TABLE 3.2. Numerical data from *Almagest* 11.10.

As for the fixed stars, Copernicus places them at an immense distance, to explain why they are not subject to any parallax, in spite of the earth's annual revolution.[21] Moreover, he postulates the immobility of the sphere that carries them, such that the greatest distance corresponds to the absence of turning, or a null period of motion.[22] The function of this stationary sphere, the container (OR, p. 21) of the universe, is to allow the moving planets it contains to be assigned precise locations: "It is unquestionably the place of the universe, to which the motion and position of all the other heavenly bodies are compared" (OR, p. 21).

Thus, when we adopt the Copernican model, "not only do their phenomena follow therefrom but also the order and size of

Chapitre troisieme des Circles de la Sphere.
B

FIGURE 3.3. The geocentric universe: Appanius, *Cosmographie* (Anvers, 1574; 1st ed. 1539).

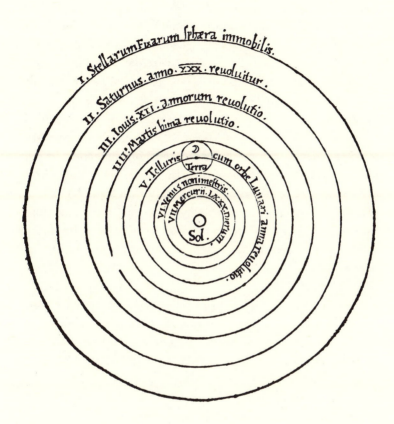

FIGURE 3.4. The heliocentric universe: Copernicus, *De revolutionibus orbium coelestium*, 2nd ed. (Basel, 1561). For Copernicus, "the dryness of numbers replaced the imagery of signs and elements. This is still not the desacrilized Universe of classical science" (Simon, *Kepler astronome astrologue*, p. 260).

all the planets and spheres" (OR, p. 5). Interpreting some epicycles and deferents as optical effects produced by the earth's
motion permits us to introduce a *standard measure* applicable to
all the planets, and to establish a general *commensurability*, leading to the recognition of a quantitative, total, and linear order.
The parts of the universe become *symmetrical* in the ancient
sense, as defined by Heron of Alexandria: "We call symmetrical the quantities that are measurable with a common standard
of measure."[23] In Rheticus' terms, "Thus there are only six
moving spheres that revolve about the sun, the center of the
universe. Their *common measure* is the great circle that carries
the earth" (3CT, pp. 146–47; my italics). The order of the
planets corresponds to increases in time and distance around
the sun. There is depth to a universe in which "the size of the
spheres is measured by the length of the time" (OR, p. 21). Or
once again:

> The first and the highest of all is the sphere of the fixed stars,
> which contains itself and everything, and is therefore immov-
> able. It is unquestionably the place of the universe, to which
> the motion and position of all the other heavenly bodies
> are compared....
>
> [The sphere of the fixed stars] is followed by the first of
> the planets, Saturn, which completes its circuit in 30 years.
> After Saturn, Jupiter accomplishes its revolution in 12 years.
> Then Mars revolves in 2 years. The annual revolution takes the
> series' fourth place, which contains the earth...together with
> the lunar sphere as an epicycle. In the fifth place Venus returns
> in 9 months. Lastly, the sixth place is held by Mercury, which
> revolves in a period of 80 days.
>
> At rest, however, in the middle of everything is the sun.[24]
> (OR, pp. 21–22)

The underlying motivation of *De revolutionibus* comes from the conviction that this "symmetry," which confers total unity on the universe, constitutes the "principal consideration," the "first principle" of the universe. It arises from an *organistic* requirement, which gives preference to "the manner in which observations at first apparently unconnected turn out to be closely related" and "conceives the value of our knowledge as proportional to the degree of integration it has attained...."[25]

*Imperfections and Uncertainties*
Copernicus divided his book into two major parts:

> Accordingly, in the arrangement of the volume too I have adopted the following order. In the first book I set forth the entire distribution of the spheres together with the motions which I attribute to the earth, so that this book contains, as it were, the general structure of the universe. Then in the remaining books I correlate the motions of the other planets and of all the spheres with the movement of the earth so that I may thereby determine to what extent the motions and appearances of the other planets and spheres can be saved if they are correlated to the earth's motions. (OR, p. 5)

The first lines of the manuscript version of Book Two, suppressed in the printed text, recall that partition: "Things that in the philosophy of nature, seemed necessary for our work in terms of principles and hypotheses...we have summarily passed in review...."[26] The beginning of Book Five does the same:

> Now I tackle the motions of the five planets. The order and size of their spheres are connected with remarkable agreement

89

and precise symmetry by the earth's motion, as I indicated generally in Book One [ch. 9]....(OR, p. 227)

The first part is anchored in the "philosophy of nature." It is based on the cosmological premise that makes of the universe an ordered whole, endowed with "remarkable agreement and precise symmetry." Its subject matter is global; it examines the acceptability of the earth's motion in terms of the "general structure of the universe." Copernicus entertains questions like the shape of the universe and the earth, their relative dimensions, and the mathematical principles that regulate the motion of celestial bodies. In chapters 7 and 8, he covers the arguments of physical science for and against the earth's motion. Then he describes the structure of the entire heliocentric universe, and "in a summary manner [*in summa*]" (OR, p. 22) the three motions of the earth implied by theory, concluding Book One with a discussion of general trigonometric notions.

In sum, Book One explores the thesis that the earth moves in terms of the systematic qualities of the universe as a whole. Beginning with Book Two, the perspective changes. According to Copernicus, he "proceeds from a general account of the earth's three motions...to explain all the phenomena of the heavenly bodies...one by one." He recomposes, for each part of the universe, the individual motions that characterize it by relating them to a moving earth. One by one, the parts of the universe are passed in review. The motion of each planet is considered as the local realization of a common regularity, namely the law of circular and uniform motion.

This detailed approach does not confirm the organistic methodology to the point that it can remove all uncertainty and eliminate all imperfection in the representation of the universe.[27]

So, although the sun is at the center of the universe, it is not situated at the (moving) center of planetary motion. This center corresponds to the center of the earth's sphere, which does not coincide with the location of the sun. The Copernican universe is thus not truly a universe with a unique center. That this phenomenon could only have been felt as an imperfection is illustrated by Copernicus' critique of the multiplicity of centers in the geocentric model.[28]

Another point concerns the *precise* relationship between the distance of a planet and the period of its revolution. Koyré notes that the Copernican model, while in complete agreement with the principle of interrelations, remained nonetheless unsatisfactory: "the proposed explanation, which implies that the planets move with equal (orbital) velocities, and consequently that the motive forces responsible for their motions are similarly equal, does not agree with the facts: the periods of revolution are not directly proportional to the distances, but differ considerably therefrom...."[29]

If Copernicus criticized the uncertainty of geocentric hypotheses for allowing somewhat different representations of the same movements (OR, p. 4), he also proposed geometrically divergent constructions without being able to eliminate one in favor of the other. Thus, he "saved" the appearance of the inequality of the sun's motion in three different manners, acknowledging that "it is not easy to decide which of them exists in the heavens."[30] Elsewhere (5.32), he proposed two equivalent theories for Mercury's longitudinal motion, specifying that the second is neither "less plausible" nor "less reasonable" than the first.

Much has been written on the utility of Copernicus' geometrical constructions. It is generally conceded that his representations are scarcely less complex or more economical than Ptolemy's, or their predictive accuracy greater.[31] Moreover, some defects in Ptolemaic theory, criticized by Rheticus and others, do not disappear with Copernicus: the problem concerning the phases of Venus, for example, arises in practically the same way.[32]

For Copernicus, all this was not sufficient to place in doubt the superiority of his theory. Obviously of greatest importance for him was the "symmetrical" *system* derived from his organistic perspective and absent from the Ptolemaic representation.[33] It is sufficient that the organistic model is not invalidated. That imperfections or doubts continue is of lesser importance: they also exist for the alternative, the hypothesis of a stationary earth, which moreover does not account for the systematic nature of the whole.

### "Symmetry's" Intertext

*From the Intratextual to the Intertextual*
For Copernicus, the organistic affirmation that the universe is symmetrical functions as a presupposition, an *a priori* framework that orients research and defines the "principal" quality that a true representation of the universe must possess. It is not part of the empirical or analytical content. It belongs rather to the global framework in which the recording and analytical processing of empirical data takes place.

Clearly the presuppositional function of Copernicus' concept of "symmetry" is based on his opposition to Aristotle's conception of the role played by the proportionality of time and distance. The *De caelo* (2.10) explains symmetry in terms of the action of the sphere of fixed stars on the other spheres. According to Aristotle, the sphere of the fixed stars moves with a simple and rapid motion in the opposite direction to the rotation of the other spheres. The sphere closest to the fixed sphere moves the slowest; the speed of successive spheres increases in proportion to their distance from the fixed sphere:

For it is the nearest body which is most strongly influenced, and the most remote, by reason of its distance, which is least affected, the influence on the intermediate bodies varying, as the mathematicians show, with their distance.[34]

For Aristotle, the "symmetry" of time and distance is the result of a preexisting state of affairs: the physical relationship between the motion of the external sphere and the motions of the spheres contained within it. At a stage in empirical analysis, symmetry comes into play, but it hardly constitutes an *a priori* framework. Far from orienting a research program or furnishing it with a framework in which to make sense, symmetry is one discrete element in a linear progression – an element, moreover, surrounded by some confusion.[35] For Copernicus, on the other hand, the sphere of fixed stars is stationary. "Symmetry" is not the effect of a prior physical relation, but an immediate end, the "principal consideration," the "first principle" of Creation. Thus it requires a reorganization of the overall picture of the universe and the elaboration of a new physical theory. And so Copernicus returned to the Pythagorean-Platonic conception of a cosmological order founded on quantity and proportionality.

Is it by chance that in the *De revolutionibus* the return to a quantitative foundation is immediately illustrated with a reference to art? Doubtless because of a desire to establish a "blank" rhetoric, a "writing degree zero," as the normative ideal of scientific discourse – linked only to a perfectly neutral and transparent metalanguage[36] – we tend sometimes to minimize the importance of this comparison, to reduce it to the level of an analogy with a purely didactic or suasive function.[37] In the sixteenth century, it was nonetheless taken seriously, most notably by Tycho Brahe, who discussed at length the reality of Copernican "symmetry" and explicitly referred to "the research of Dürer, this excellent painter, on the symmetry of the human body, which is the Microcosm."[38]

In any case, the comparison only underlines what we could also establish without it, with the help of other passages: namely, Copernicus' insistence on the *theme* it illustrates ("principal consideration," "first principle," "marvelous symmetry") – an intratextual insistence by means of which Copernicus' text communicates objectively with the Renaissance *intertext*. It is significant, for example, that Erasmus' praise of Dürer includes this formula: "He observes symmetries and harmonies exactly" [*Observat exacte symmetrias et harmonias*].[39] Whether there is direct influence or not, for Copernicus the signified – "God" – is available to be read and is bound to a semantic field for which Renaissance art provides an exemplary illustration. Both refer to the same complex of representations, projected as much on God as on the artist, so much so that we can see reciprocal metaphors: if the Renaissance artist is often called a god,[40] Copernicus' God creates like a Renaissance artist.

*Return to Polykleitos*

The concept of *symmetry* posed a terminological problem. In his *De sculptura*, Pomponius Gauricus noted that there is no exact Latin equivalent for the Greek term: he proposed *commensuratio*, which Cicero had already adopted and which accurately rendered the etymological sense of *symmetria*.[41] Alberti had used *concinnitas*, stressing the primordial significance of the concept.[42] Subsequently, other terms appeared. Equicola referred to *proportione et conveniente quantità*.[43] Later Lomazzo used *proportione, consonanza, rispondenza, commodulatione*.[44] Dürer introduced *Vergleichlichkeit*.[45] Baldus' Vitruvian lexicon listed as synonyms *commensus, proportio, correspondentia, rechtmessung, gleichformung*.[46] In French, Blaise de Vigénère proposed a circumlocution: *la deuë convenance des proportions* (proper conformity of proportions),[47] but Simon Goulart made do with *proportion*, just as Perrault would in the seventeenth century.[48] These samples are far from exhausting all the terminological solutions that were proposed, but their very proliferation gives some idea of the fascination this idea exercised, and its omnipresence.

For the entire Renaissance, whatever translation was used, Vitruvius is the point of reference for the aesthetic ideal we are discussing. According to Vitruvius,

> Symmetry is the proper agreement between the members of the work itself, and relation between the different parts and the whole general scheme, in accordance with a certain part selected as a standard. Thus in the human body there is a kind of symmetrical harmony between forearm, foot, palm, finger, and other small parts; and so it is with perfect buildings.[49]

In this passage, as in the following and in Copernicus, the notion of *symmetry* relates first and foremost to the body and its parts:

> The design of a temple depends on symmetry, the principles of which must be most carefully observed by the architect. They are due to proportion, in Greek *analogia*. Proportion is a correspondence among the measures of the members of an entire work, and of the whole to a certain part selected as standard. From this result the principles of symmetry. Without symmetry and proportion there can be no principles in the design of any temple; that is, if there is no precise relation between its members, as is the case of those of a well-shaped man.[50]

Looking back beyond Vitruvius, this aesthetic ideal is derived

95

from the Pythagorean tradition. A principle of delimitation and of ordering that defines the distinction between cosmos and chaos, number is also the law of beauty. The oldest known theoretical formulation of the aesthetic of *symmetry*, is contained in Polykleitos' *Canon*, especially as passed on by Galen:

> Chrysippus maintained that beauty does not consist in the elements, but in the symmetry of the parts, the relationship of one finger to another, of all the fingers to the rest of the hand, of the rest of the hand to the wrist, of the wrist to the forearm, of the forearm to the entire arm, and of each part to every other part, as it is written in the Polykleitos' *Canon*. And Polykleitos, who taught us the proportions of the body in this document, corroborated his theory by action, since he created a statue conforming to the precepts of his theory, and he called that statue *Canon*, the same title as his written work.[51]

Panofsky has pointed out the strong similarity between the aesthetics of Polykleitos and of the Renaissance, of Dürer with his demand for "agreement of one with the other" [*Vergleichlichkeit Eines gegen dem Andern*].[52] For Dürer as for Copernicus, the aim is not to "save phenomena," to use the traditional astronomer's expression, nor to propose an "expeditious method" conceived "to work lightly" [*pour legierement ouvrier*], in the terms Villard de Honnecourt applied to artists. In one sense, moreover, Villard (plate 3) is to Dürer (plate 4) as Ptolemy to Copernicus: he represents the "mechanism" that models the body by juxtaposing the parts – which themselves are combined from smaller units – without any concern for the "symmetry" of an organic unit.[53] For both Copernicus and Dürer, the major objective was to discover the principle informing Creation – Creation of the heavens as well as man, since the organic unity of both is based on number. When Copernicus set up the earth's sphere as the "common measure" that alone can determine the proportions of size and length for all the other spheres, he is applying Dürer's rule for artists:

> Above all, we must seek *the most certain and suitable measure of all the parts.* Having established this, by following its order and manner, we can bring to light with great care and diligence all the parts, large and small, if by chance we can find some beauty and approach perfection.[54]

The parallel with Renaissance aesthetics is all the stronger, given the profusion of counterexamples produced by the theoreticians of art, where the result of the absence of "symmetry" is, in Copernicus' terminology, *monstrous.* Here is Leonardo da Vinci:

> it would please me if you flee monstrous things, such as long legs combined with a short torso, or narrow chests and long arms....[55]

Alberti defined the ideal as follows, also adding some counterexamples:

> In the composition of the members, it is necessary in the first place to be sure that they all agree with each other.... For if in a picture the head is very large, the chest very small, the hand very wide, the foot puffed up, the body swollen, this composition will certainly be ugly to look at. One must observe conformity in the size of the members and in this submission to a common measure....[56]

Monstrosity arises from a lack of proportion between the parts

97

FIGURE 3.5. Mechanist geometrism: Villard de Honnecourt (seventeenth century).

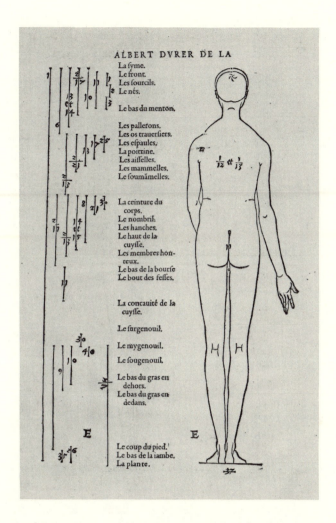

FIGURE 3.6. Organicist "symmetry": Albrecht Dürer, *Quatre livres de la proportion des parties et pourtraits des corps humains* (Paris, 1557; 1st German ed. [*Vier Bücher von Menschlicher Proportion*] 1528). Villard de Honnecourt's mechanist mathematicization is to Dürer's organicist mathematicization as Ptolemy's system is to Copernicus'. See pp. 105–06.

of an organic body: the counterexamples reappear with great detail in Dürer:

> The entire image, from head to foot will be in proportion, in whatever mode is in order, whether of a strong person or weak, fleshy or thin, so that one part does not appear too fleshy and another too gaunt: for example the arms are thin and the legs fat; or the figure is full from the front and not from the back, or the reverse; we truly want the parts to be proportioned among each other, and not badly thrown together, and without reason.[57]

The "principal consideration" that led Copernicus to turn away from geocentric theories thus appears as a theme not limited to Renaissance science. As has often been noted, from the point of view of its agreement with the empirical data or its predictive utility, the Copernican system presented no real advantage over its geocentric rivals. Its superiority resides chiefly in the coherence it introduced to planetary motion. No doubt this coherence is eminently significant from the point of view of scientific logic, but the fact remains that Copernicus himself presented it in an aesthetic light. He insisted moreover that thanks to his system "heaven itself is so linked together that in no portion of it can anything be shifted without disrupting the remaining parts and the universe as a whole" (OR, p. 5). The affinity between this kind of declaration and the principle articulated by Alberti in *De pictura* as well as *De re aedificatoria* is too striking for us not to accept that the same presupposition is at work:

> Having selected this one member, the rest should be accommodated to it, so that there is no member of the whole

body that does not correspond with the others in length and breadth.[58]

But as in an Animal, the Head, the Feet, and every particular Member, should be exactly proportioned to all the other Members, and to all the rest of the Body; so in a Building, and especially in a Temple, all the Parts should be made to correspond so exactly, that let us consider which of them we please, it may bear its just Proportion to all the Rest.[59]

*The Anamorphosis of the "Symmetrical" Body*
The search for "symmetry" and the adoption of the geometrical perspective share the same organistic objective. Both assign value to the whole based on the mathematical proportions between all its elements. According to Piero della Francesca, "as the word implies, perspective refers to things seen from afar, represented proportionately within specific conditions *according to their distances*" [la prospettiva sona nel nome suo como dire cose vedute da lungi, representate socto certi data termini con proportione, *secondo la quantità de le distantie* loro...].[60] For Leonardo da Vinci, perspective introduced musical harmony: objects placed at regular intervals "diminish" according to a regular progression.[61] In the same spirit, Jean Pélerin wrote that "[q]uantities and distances have harmonious differences."[62] What is always important is proportion, harmony. The focus shifts from consideration of the parts of an object in itself to the distance between subject and object. Submitting the representation of an object to a mathematical rationality based on its relation to the subject confers a supplementary meaning to the formula that man is the *measure* of all things.[63]

Copernicus combined the concept of a "symmetrical" Creation with the requirements of an organistic representation based on mathematical reason. If the "symmetrical body" is the model for Creation, the most faithful representation is necessarily the one that reproduces its immanent proportions. Although man has access only to his own representations, he can discriminate among these: the representation that captures the "symmetry" of the universe will be the true one.

Such a representation is possible, however, only if we conceive of man's viewpoint in an uncustomary manner. If the astronomers who adopt geocentric systems cannot capture the coherence and the "symmetry" of the universe, it is because they fail to take into account the true nature of their viewpoint. God has created a "symmetrical" universe, but this "symmetry" only appears to man if he assumes that he is not viewing things from the center. The Copernican revolution accomplished a kind of anamorphosis of the universe, since it required man to abandon his customary point of view in favor of one that is uncustomary but appropriate to viewing God's Work. When he compared the current representation of the universe to a monster, Copernicus implied a kind of disproportion and disfigurement that also applies to pictorial anamorphosis when the correct point of view has not been discovered. Thus Jean François Nicéron declared that he had revealed "the means of constructing several sorts of figures from a frontal perspective such that they appear deformed and nonsensical when seen from a different point of view, yet appear well proportioned when seen from the correct point of view."[64] Leurechon proposed the following problem: "Disguise a figure, such as a head, an arm, or an entire body such that it is out of proportion...and, nonetheless, seen from a certain point it will appear in correct proportion."[65] Like an anamorphosis not seen from the correct vantage point, the geocentric universe displays a distorted body with disproportionate members, thus preventing us from representing it as a unified whole. Artistic anamorphosis, we should remember, is based not on real objects but on an existing representational technique with a perpendicular point of view. Its object is not the thing represented but a representation that it transforms: it moves from cipher to counter-cipher, *cifra e contracifra*.[66] The heliocentric model is arrived at by a controlled transformation of the existing geocentric model. Copernicus employed Ptolemy's technical machinery (eccentrics, epicycles), but reordered around a center that is not the "center of terrestrial gravity" (OR, p. 118).

The truth of Copernicus' anamorphosis of the world depends entirely on its effect, which responds to the *a priori* demand for a universe governed by *certa ratio*. Likewise for Alberti, the *certa ratio* determines the validity of pictorial perspective by enabling "knowledge by comparison" of the elements in the representation.[67] Copernicus and Alberti focus attention on the relational form of *both* the representation and what it represents. For Copernicus as for Renaissance painters, the "principal consideration" is the immanent causality of a system of reciprocal relations, one law ruling the formal disposition of the parts of the universe in a representation *"distinct from all matter"* [*ab omni materia separata*], to use the terms employed by Filelfo and Alberti.[68] This is an Apollonian universe, where "the idea or substance of a thing is merely a position in space...determined precisely by the fact that it gives a situation proportionately (*per comparatione*) to all the accidents, that is to say, because it reabsorbs and eliminates the matter of which the thing is composed into a system of proportional relations."[69] These words, with which Giulio Carlo Argan describes Brunelleschi's work, also apply to Copernicus' universe, governed by quantitative relations in homogeneous space, and not by distinctions between elements or material objects situated in qualitatively differentiated, distinct locations.

# CHAPTER FOUR

# Synecdoches

Copernicus replaced Ptolemaic mechanism with organistic "symmetry," but to explain motion he maintained – and even reinforced – the formism of antiquity. Motion in the heavens is the teleological realization of circular perfection. The circle and sphere are endowed with ontological dignity. Through circular motion, the spherical body of a planet "expresses its form as the simplest body, wherein neither beginning nor end can be found, nor can the one be distinguished from the other" (OR, p. 10). Apparent irregularities in planetary movements must be understood as illusory phenomena, which science is obliged to eliminate for reasons that are accepted *a priori*:

> For this nonuniformity would have to be caused either by an inconstancy, whether imposed from without or generated from within, in the moving force or by an alteration in the revolving body. From either alternative, however, the intellect shrinks. It is improper to conceive any such defect in objects constituted in the best order. It stands to reason, therefore, that their uniform motions appear nonuniform to us. (OR, p. 11)

Concretely, the formistic approach exacted, among other things, three "transformations" of the circle and sphere concerning respectively physics, the elimination of the equant, and the anagogical role of the sun's centrality. Because of its magnitude, I will take up the last point in a separate chapter, but it is important to recognize that these three innovations clearly correspond to operations described in rhetoric as tropes. In the following chapter I will discuss metaphor, in this chapter the two principal relationships that synecdoche brings into play: the whole and the part, the genus and the species.

## The Whole and the Parts

### Order and Unity in Physics

Some contemporary objections to the earth's motion were based on Aristotelian physics, which distinguished supralunary from sublunary motion, associating each with specific elements. Celestial matter, or ether, moves naturally in a circle around the center of the universe, while lighter elements of the sublunary world (fire and air) tend naturally to move upward (from the center), and heavy elements (water, earth) downward (toward the center). Gravity is the propensity of heavy objects to move to the center of the universe, which coincides with the center of the earth. Because of its heaviness, the earth naturally seeks a state of rest at the "lowest" place. As Copernicus reminds us, "Earth is in fact the heaviest element.... All the more, then, will the entire earth be at rest in the middle, and as the recipient of every falling body it will remain motionless thanks to its weight" (OR, p. 14).

If the earth's central position can be explained by a natural tendency arising from its heaviness, we must add that each element, according to Aristotle, has only one natural motion.[1] If the earth's motion were circular, this, in combination with a natural motion toward the center, would produce a necessarily *violent* motion, and like all violent motions, would be incapable of persisting. Moreover, according to an argument also mentioned by Copernicus, this violent motion would be of such rapidity that it would inevitably lead to the earth's explosion: "But things which undergo an abrupt rotation seem utterly unsuited to gather [bodies to themselves], and seem more likely, if they have been produced by combination, to fly apart unless they are held together by some bond" (OR, p. 15). At the very least, the violence and speed of the earth's motion would produce effects in direct opposition to those we experience. How are we to explain the fact that birds and clouds are able to move from west to east? And an object in free fall from a tower, representing rectilinear motion toward the center of the universe, would never strike the earth's surface at the foot of the tower if the earth were itself in motion. Du Bartas collected a number of these examples in his refutation of the Copernican system:

> So, never should an Arrow Shot upright,
> In the same place upon the shooter light:
> But would doo (rather) as at Sea, a stone
> Aboord a Ship upward uprightly throwne,
> Which not within-boord falles, but in the Flood
> A-stern the Ship, if so the wind be good.
> So, should the Foules that take their nimble flight
> From Westerne Marshes toward *Mornings* Light
> and *Zephirus*, that in the Summer-time
> Delights to visit *Eurus* in his clime,
> And Bullets thund'red from the Canons throat,
> (Whose roaring drownes the Heav'nly thunders note)
> Should seeme recoyle: sithens the quicke careere,
> That our round Earth should daily gallop heere,

Must needs exceed a hundred-fold for swift,
Birds, Bullets, Winds; their winds, their force, their drift.[2]

Elsewhere, some of these arguments were articulated with greater subtlety. According to a remark by Tycho Brahe, a cannonball shot toward the west should strike a more distant target than one shot toward the east, since the earth's motion would bring the projectile and the target toward each other in one case and away from each other in the other.[3]

To explain the effective motion of terrestrial bodies, to account for the absence of an atmospheric hurricane or even an explosion induced by the earth's motion, Copernicus introduced another physical theory. In the first place, he postulated that if the earth does move, its motion is *natural*, and therefore cannot produce *violent* effects that would lead to a hurricane or explosion:

> Yet if anyone believes that the earth rotates, surely he will hold that its motion is natural, not violent. But what is in accordance with nature produces effects contrary to those resulting from violence, since things to which force or violence is applied must disintegrate and cannot long endure. On the other hand, that which is brought into existence by nature is well-ordered and preserved in its best state. (OR, p. 15)

At the same time, Copernicus defined gravity as the tendency of the parts of a whole to maintain their unity in a spherical shape instead of a natural tendency toward the absolute center of the universe: "For my part I believe that gravity is nothing but a certain natural desire, which the divine providence of the Creator of all things has implanted in parts, to gather as a unity and a whole by combining in the form of a globe" (OR, p. 18). If the

108

earth is characterized by eccentricity and motion, everything possessing "earthiness" will naturally follow the circular motion of the whole to which it belongs. "Therefore rectilinear motion occurs only to things that are not in proper condition and are not in complete accord with their nature, when they are separated from their whole and forsake its unity" (OR, p. 17). Rising and falling, the motion of the parts of a whole is double, at once circular (accompanying the whole) and rectilinear (from or toward the center of the whole).

Thus, rather than ordering motions in terms of absolute locations, as did Aristotle, Copernicus relates them to as many centers as there are wholes:

Each such motion encircles *its own center:*

*This* [fact that the earth is not the center of all revolutions] is indicated by the planets' apparent nonuniform motion and their varying distances from the earth. These phenomena cannot be explained by circles concentric with the earth. Therefore, *since there are many centers*, it will not be by accident that the further question arises whether the center of the universe is identical with the center of terrestrial gravity or with some other point.... This impulse [Copernican gravity] is present, we may suppose, also in the sun, the moon, and the other brilliant planets, so that throughout its operation *they remain in that spherical shape which they display. Nevertheless, they swing round their circuits in divers ways* (OR, pp. 17-18).

In contrast to Aristotle, Copernicus attributed in effect a variable reference to the idea of "physical totality." For Aristotle the universe as a whole maintains its physical "integrity." Copernicus transfers this property to the sun and planets, each of which he treats as a totality. From the physical point of view, Copernicus sets up an internal relation of equivalence between the parts of the universe, thereby completing the internal relation of the order of the planets established by the mathematical ratio between distance and period of orbit. Insofar as he attributes to the parts what hitherto only applied to the whole universe, his physical theory is the result of a synecdoche, and more particularly a particularizing synecdoche.[4] A multiplicity of "centers" take the place of a unique center; the parts take the place of the Aristotelian whole and themselves become totalities with a tendency to take on a spherical shape:

> I mean the sun, moon, planets, and stars, are seen to be of this shape; or that wholes strive to be circumscribed by this boundary, as is apparent in drops of water and other fluid bodies when they seek to be self-contained. (OR, p. 8)

The consequence of this synecdoche is that physics can no longer give us any assurance concerning the global arrangement of the universe, a task which now falls to mathematics.

According to the Aristotelian scheme, if the earth were carried away from the center, "the various fragments of earth would each move not toward it but to the place in which it now is, that is to the absolute center of the universe."[5] For Aristotle, space, "exerts a certain influence," and "places do not differ merely in relative position, but also as possessing distinct potencies."[6] Space plays an active role in the movement of things that pass through it; each element is attracted by its proper place. One important consequence of the Aristotelian point of view is that mathematical astronomy is secondary to physical astronomy; its role can be "descriptive but not explanatory."[7] Physics analyzes the places that make up the cosmos, explains their differences and their place in the global order. Mathematical astronomy must conform to the physical framework, compared to which its constructions appear increasingly hypothetical, even fictitious.

Copernican physics, on the other hand, limits itself to the description of local systems. Every "fragment" of earth seeks its "own center." Physical laws are no longer linked to qualitatively differentiated space. There is no opposition between the sublunary world and the supralunary; the same principle governs both. Suddenly, the relation between physics and mathematics is reversed. Physics cannot ascertain the global organization of the universe, the "cosmicality'" of the cosmos. For this, we must turn to mathematics, in conformity with the fundamental theme of "symmetry" that we touched on in the preceding chapter.[8]

### Neoplatonic Connections

Copernican physics was doubtless to some degree dependent on medieval physics. The arguments and examples had been prefigured by Buridan and Oresme.[9] Duhem has suggested the influence of Nicholas of Cusa, who had already rejected the idea of the particularity of place in favor of homogeneous and continuous space, and had reflected on the rotation of the sphere based solely on the sphere's geometrical properties.[10] But the differences are equally important. The cardinal denied the possibility of a mathematical study of nature; his universe was indeterminate and lacked a center; the sun also moved, and so on.[11]

Pierre Kerszberg has looked for similarities with Plato. The *Timaeus* describes the propulsive function of the world soul as expansion from the center. Circular motion is conceived as the most appropriate form of motion for this soul, which has no determined physical location, but everywhere radiates outward, and generates the order and mathematical harmony of celestial bodies.[12]

111

Z. Horsky has emphasized that it is more appropriate to make comparisons with Marsilio Ficino's Neoplatonic texts than with Plato himself.[13] Horsky's point is worth dwelling on for a moment. Even though Ficino's thinking always remained bound to a geocentric conception of the universe and was often expressed in Aristotelian terms,[14] a number of its elements prefigure characteristics of Copernican physics. I am not trying to establish Ficino as the model or source of Copernican physics, only to indicate tendencies and common patterns of thinking.

Ficino consistently asserted that circular motion is natural to all the elements. And while he believed that the earth is stationary and at the center of the universe, he nonetheless argued that if the earth did move, its motion would be circular:

> Fire and air turn in a circle, like the moon, as the motion of the comets demonstrates. Water turns in a circle, flowing back without interruption. If the earth moved, as Higesias thought, its motion would be circular, although according to the opinion of most it is at rest.[15]

This hypothesis asks us to imagine the possible motion of the earth as *naturally* circular (as Copernicus argued); conspicuously absent is any mention of the violent and destructive effects that according to Aristotelian physics would inevitably accompany the earth's circular motion.

Ficino also tended to disregard the distinction between sublunary and supralunary regions. He attributed a soul to each celestial sphere as well as to the four terrestrial elements:

> Air has its soul, fire its, for the same reason that earth and water have theirs. Likewise, the eight heavenly spheres have eight souls, for the ancients had eight heavens.[16]

*Likewise* (*similiter*): the consideration of likeness, allowing only differences of degree, tends to replace the radical Aristotelian distinction between the sublunary and the supralunary.[17] Ficino insists, moreover, on the specifically Platonic character of attributing life to the elements: "But if for the Platonists it is absolutely obvious that the spheres of the elements live, the ancient Peripatetics did not treat this problem."[18] The language with which Copernicus speaks of motion is tinged with animism and could indicate a Neoplatonic approach.[19] Copernicus' typically terse ascription of life to the earth, which "has intercourse with the sun, and is impregnated for its yearly parturition" (OR, p. 22), recalls Ficino's amplification of the theme in *Theologica platonica*:

> We see the Earth give birth, thanks to varieties of seeds, to a multitude of trees and animals, nourish them, and make them grow; we see her cause even stones to grow as her teeth, vegetable life as hairs, as long as they remain connected to their roots, while if they are removed or unearthed, they cease growing. Could we say that the breast of this female lacks life, she who spontaneously gives birth and nourishes so many offspring, who sustains herself and whose back carries teeth and hair?[20]

We see that Ficino, like Copernicus, distanced himself from the Aristotelian theory of attraction in favor of empty and abstract space. Whereas Copernicus conceived of gravity as the natural tendency of parts to coalesce into their specific wholes (the earth, the moon, etc.), Ficino reasoned in terms of the attraction proper to the soul of each sphere: "And, if we ask about the rise and fall of elements and composite objects, the Platonists reply that they are accomplished by the soul, I mean by the soul of their sphere, which, just as a magnet attracts iron, attracts the parts of their sphere."[21] Here also, we find a synecdoche at work and the abandonment of the notion that abstract places exercise individual forces.

*Art: The Circle*

If such points of likeness exist between Copernican physics and the rebirth of Neoplatonism as represented by Ficino, we have yet to mention Ficino's insistence on "nature the artist." He compared nature to "the intelligence of the geometrician when he inwardly constructs imaginary matter."[22] Art and geometry: Koyré also discerns in Copernican physics an "aesthetics of geometry."[23] Need we recall that Renaissance art generally sought to unify the representation of the human body in terms of the circle, that with the Vitruvian drawings of Leonardo da Vinci, Dürer, and others, a kind of grammar was sometimes constructed, according to which the parts of the body were subjected, as they completed the most varied movements, to circular units of measurement? The *Codex Huyghens*, a sixteenth-century Italian manuscript inspired by Leonardo's (lost) *Libro del moto actionale*, includes striking illustrations of an analysis of the motions of the human body, using a system of circles and epicycles; it makes the analogy with the motion of celestial bodies explicit, since the author declared that he was proceeding, "according to our primary order of the motion of the celestial bodies" [*sequendo il primo nostro hordine del moto de' corpi celesti*].[24] Leonardo da Vinci represented the growth of the branches of a tree as an expansionary motion in concentric circles around a geometrical point: change must adhere to the realization or maintenance of a formal whole, of a unified *eidos*. To this circular measure, he added, moreover, a principle of commensurability or "symmetry." Along any point on the circumference, the sum of the thicknesses of all the branches must equal the thickness of the

114

trunk.[25] This brings to mind Raphael's paintings, where

> the composition is organized in terms of a circle into which
> the human figures must fit; the circle is an emblematic form;
> it reflects unity of order.... Beyond mere ease of gesture and
> liberty of line, each character in the painting rigorously fits
> within this geometry and submits to it effortlessly, *naturally*,
> is witness to the harmonious motion that the circle inevita-
> bly produces and that confirms its theological meaning."[26]

In spite of this *imagination* of a geometrically autotelic circle, it
is true that art theory continued to *conceptualize* motion in an
essentially Aristotelian fashion.[27] In Leonardo's texts on astronomy, however, we encounter some fragments that employ a synecdoche similar to the one that characterizes Copernicus' physics.
Sometimes Leonardo conceived of the universe as containing
homogeneous space, with multiple centers of attraction replacing
an absolute center. The earth is not the center of the universe,
but of its own elements.[28] The moon also contains water, air,
and fire; these are drawn to their own center, just as terrestrial
elements are drawn to theirs.[29] But elsewhere Leonardo maintained a conception of space in which direction and absolute
location are significant. When he hypothesized that the earth was
removed to a point distant from the center of the universe, he
argued that water would continue to form a sphere around the
center of the universe, rather than around the earth....[30] Such
thinking is not coherent, but one direction in which it is drawn
is toward Copernicus.

*Art:* Perspectiva Superior
Once the distinction between physical and metaphysical regions
had been erased, the Copernican universe was free to imply "the

geometrization of space – that is, the substitution of the homogeneous and abstract space of Euclidian geometry for a qualitatively differentiated and concrete cosmic space."[31] This space is like the perspective space of Renaissance painting, which in principle is also continuous and homogeneous; every part is subject to the same laws. Moreover, the question of the infinity of space tormented discussions of artistic perspective beginning with Alberti (1404–1472) as well as cosmological speculations beginning with Nicholas of Cusa (1401–1464). Giorgio de Santillana has emphasized the relation between the optical vanishing point and Nicholas of Cusa's notion of *complicatio*:

> We are asked to see longitudinal perspective in terms of that other inverted pyramid ending in the flight-point placed on the infinite circle: that mathematical point at infinity in which all the forms and ratios of reality are absorbed or rather "contracted." No one could deny that we are here on Cusan territory, although there could be no question of direct influence.... Alberti is fully aware that the longitudinal perspective carries the theme of contemplation; it loses itself as Plotinus suggests in that one flight-point. In fact, it is the equivalent of the medieval golden background: But the imagined ensemble of flight points is, with respect to that sensuous gold, utterly abstract – a true intellectual construction, like that of Nicholas of Cusa, ending up on the "circle of infinite radius."[32]

In the homogeneous space of the Copernican universe, every place is in principle a possible center. No law predisposes a particular body to be in one place rather than another, since physical law only determines the spherical unity of individual bodies. Infinite *space* without differentiated locations is nevertheless not the *cosmos* that Copernicus postulated. If each motion has its own center, the created universe also possesses an absolute center, which distinguishes it from the homogeneous space of pure potentiality: "We will finally grant that the sun itself occupies the center of the universe."[33] The universe may no longer be organized in terms of different substances, each provided with its own proper location, but it is organized according to the principle of "symmetry" as it relates to this center. Two grand mathematical qualities are assigned not to space, but to Creation – possession of a privileged center and "symmetry" of distance and motion. These confer a quality on the Copernican cosmos that also characterizes how human figures are inscribed in the universe of the artistic work.

For Brunelleschi and Leonardo, there already existed what Guido Argan has called "dramatic" perspective[34] and what Robert Klein characterizes as "compositional perspective."[35] In *De sculptura*, Gauricus called it *perspectiva superior*.[36] In effect, the artist must do more than inscribe figures in a given space. He must construct a space that is adaptable to the requirements of *history*: sufficient intervals between actors, coordination of the number of actors and the "rational" filling of space, the effective staging that gives *istoria* duration by referring to a before and an after, and so on. This other "perspective" is no longer tied to optics, but to poetics (the demand for clarity, verisimilitude, unity) and rhetoric (Gauricus constantly borrows from Quintilian).[37]

*Perspectiva superior* is present in Copernicus. It is true that space in the Copernican universe is homogeneous and that although *De revolutionibus* does not speak of the infinite extension of the universe, at least it calls the universe "immense" (1.6). But the cosmos that is inscribed in that homogeneous space is still provided with an absolute center, an anchorage point for the "symmetry" of motion. Copernicus' enthusiasm for the sun's centrality reveals his desire to protect created space from the homogeneity of Euclidian space where all locations are like all others. The sun's place arises from the distinction (*eukrineia*) that Gauricus presented as one of the two principal aspects of *perspectiva superior* and "which means that a thing pleases by its adornment and suitability."[38] For Copernicus, the sun's location is eminently *suitable*: it is not located at a random point in infinite space but at the predominant place within created space. We are reminded of Lomazzo's following lines: "The universal perspective shows how to arrange an isolated figure in terms of placement and necessary accompaniment, such as arranging a King in a posture of majesty that suits his condition, on a raised and predominant place...."[39]

The other quality of *perspectiva superior* lies – still according to Gauricus – in the *purity* (*sapheneia*) that results "when the clarity of an image impresses itself upon the eyes of a spectator."[40] This clarity is comparable to Copernican "symmetry," which replaces the uncertainty of the Ptolemaic composition of the universe, in which it is impossible to decide the exact placement of some of the planets, with a clear and pure ordering by time and distance.

The Copernican universe is distant from homogeneous space. The cosmos obeys something like a compositional perspective, which also in art supplements the geometrization of space.

### The Genus and the Species

In Ptolemaic astronomy, there is no commensurability between the distances of the planets and the periods of their orbits. Furthermore, the motion of each planet can be made to obey the law of uniformity only by recourse to a supplementary circle, the *equant*, whose center coincides with neither the center of the planet's orbit nor the center of the universe (Figure 4.1). Such a

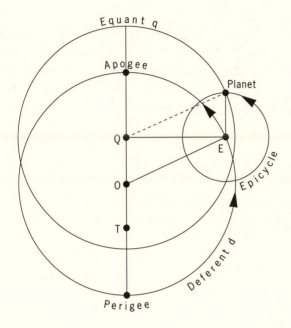

FIGURE 4.1. The equant. A planet's angular velocity is not uniform with respect to the center of its deferent (O), but with respect to the center of another circle (Q). For the outer planets and Venus, the equant point Q is on the line formed by the apsides (connecting the apogee and perigee, passing through O and T, the earth), while OQ=OT ("bisection of the eccentric"). For Mercury, Q is between the earth and the center of the circle carrying the moving deferent point. For the moon, Q is itself a moving point.

construction contradicts a principle that Ptolemy himself had articulated:

> it is first necessary to assume in general that the motions of the planets...are all regular and circular by nature.... That is, the straight lines, conceived as revolving the stars or their circles, cut off in equal times on absolutely all circumferences equal angles at the centres of each....[41]

The artifice involved in the equant circle had already been criticized in antiquity.[42] In the sixteenth century, Fracastoro, who wanted to revive homocentric astronomy, saw the equant as an "impious" invention, "unworthy of Heaven."[43] In the opening lines of his *Commentariolus*, Copernicus also expressed forcefully his dissatisfaction with this device:

> Yet the planetary theories of Ptolemy and most other astronomers, although consistent with the numerical data, seemed likewise to present no small difficulty. For these theories were not adequate unless certain *equants* were also conceived; it then appeared that a planet moved with uniform velocity neither on its deferent nor about the center of its epicycle. Hence a system of this sort seemed neither sufficiently absolute nor sufficiently pleasing to the mind.
>
> Having become aware of these defects, I often considered whether there could perhaps be found a more reasonable arrangement of circles, from which every apparent inequality would be derived and in which everything would move uniformly about its proper center, as the rule of absolute motion requires.[44] (3CT, pp. 57-58)

As he would repeat in *De revolutionibus* (4.2, 5.2), Copernicus refused to concede that the uniformity of circular motion should be measured on a *foreign, unsuitable* circle, where the motion *does not really take place*. Reducing phenomena to *regular* and *suitable* circular motion is the "principle of art." And so Copernicus considered the elimination of equants one of the principal advantages of his system. Rheticus also emphasized this point:

> Furthermore, most learned Schöner, you see that here in the case of the moon we are liberated from an equant by the assumption of this theory, which, moreover, corresponds to experience and all the observations. My teacher dispenses with equants for the other planets as well, by assigning to each of the three superior planets only one epicycle and eccentric; each of these moves uniformly about its own center, while the planet revolves on the epicycle in equal periods with the eccentric. To Venus and Mercury, however, he assigns an eccentric on an eccentric.
>
> ...[M]y teacher says that only on this theory could all the circles in the universe be satisfactorily made to revolve uniformly and regularly about their own centers, and not about other centers – an essential property of circular motion. (3CT, pp. 135 and 137)

On the first page of his copy of *De revolutionibus*, Reinhold, astronomer of Wittenberg and author of the *Prutenic Tables*, noted: "Astronomical axiom. Celestial motion is uniform and circular or composed of uniform and circular motions."[45] The *punctum aequans* [equant point, center of the equant] effectively introduced a new point of reference, thereby creating a divergence between the representation of circular motion and the representation of uniform velocity. This device represented an attempt to satisfy the formistic axiom requiring uniform velocity. The representation of the planet's orbit failed to satisfy this requirement, so a supplementary circle was added. The result was ambiguous: expectations were both satisfied and disappointed: uniform motion was obtained, but only by adding a circle invented for the purpose, that is at the price of an *ad hoc* departure from the representational system that had been worked out for the orbit.

The introduction of the equant point corresponds to a metonymic operation involving the relations among the contiguous parts of an empirical representation. This operation reduces these parts to interrelated functions without referring, as would synecdoche, to a systemic necessity. Uniform velocity with respect to an equant point is only an *effect* of the geocentric model. If one accepts the earth's immobility at the center of the universe, then each planet happens to move in an effectively uniform manner around a point (Figure 4.1) for whose location there is no physical explanation. This is an unexpected (and fundamentally undesired) empirical effect. There is no "cause" within the system or point predisposed by the system to assume this function; this follows from the principle articulated by Ptolemy in the passage cited above: the equant point is not the center of motion, which should serve as a unique reference. By a metonymic shift from what properly should be the center to this "improper" point, it is possible to save the principle of uniform motion. The representation of a planet's path is the unwanted cause of an effect which in turn becomes the reference point for the representation of uniform velocity.

With the elimination of the equant point, the Copernican system substitutes a "proper" or systemic point of reference (conforming to the principles and organic unity of the representation) for an "improper" point of reference. Henceforth, the center of uniform velocity coincides with the center of motion. Synecdoche replaces metonomy. Synecdoche "activate[s] a relation of perceived integration between two unities."[46] The tropological operation corresponding to the *living* synecdoche, neither banal in and of itself, nor banalized by usage, resides in a redistribution of the relations between wholes and parts, categories and particulars. The heliocentric model enables just such an operation: the *particular*, hitherto accounted for by the equant, is *integrated* in a *generic* representation with a unique center.

This integration of uniform velocity into a global system with a unique center earned Copernicus Tycho Brahe's highest praise:

> But in our days, Nicholas Copernicus, who not without reason is called a new Ptolemy, discovered, by his own observations, that something was missing in Ptolemy and judged that Ptolemy's hypotheses granted something that was unsuitable and contrary to the axioms of mathematics.... For this reason, he constructed, with an admirable subtlety of mind, his own hypotheses in another manner, and he revived the science of celestial motion with such success that no one before him had philosophized as precisely about the course of the stars.... For Ptolemy's hypotheses and others that were in mode represented the motions of the stars on their epicycles and eccentrics in an irregular manner with respect to the center of their circles – which is absurd – and they saved in an improper manner the regular motion of the stars by an irregularity. All that in our days is evident and known concerning the revolution of the stars was elaborated and transmitted by these two artisans, Ptolemy and Copernicus.[47]

We must retain Brahe's remark that no one had ever *philosophized* so well on the stars. If Copernicus refused the artifice of the equant and required regularity of motion to be measured on the circles of the planets, it was because these circles could not be reduced, in his eyes, to simple tools for calculation. The circle is invested with *intrinsic* value and cannot be utilized in *ad hoc* constructs that locate the planets along a periphery not related to their own center. *De revolutionibus* (4.2) explicitly poses the problem in terms of the opposition *proper-foreign*. Likewise, Rheticus, in the passage cited above, interjects the *essential property* of the circle. The significance of the true center can only be explained if we refuse to consider a center as a purely instrumental point of reference, if we grant it an ontological distinction or superiority, as did the Renaissance Neoplatonists:

> Thus the point that is traced on a line of the circle is neither on all the other lines, nor extended on the entire line or in the entire circle. But the point that is the center of the circle, which is restricted to no line, is in a certain manner in all the lines that lead from the center to the circumference. And although no point on the circumference relates equally to the entire circle, nonetheless the center, which is not fixed on a circumference, looks equally on the entire circle.[48]

Even if there is no direct link between these lines by Ficino and Copernicus' rejection of the equant, they define an attitude toward mathematics as a representational system with symbolic resonance, which respects the preeminence of the center, and where formism rejoins organicism. Only the true center is present, *contained*, *diffused* in each of its radii, without *belonging* to any one of them in particular. The center of the equant certainly does not possess these qualities. Its relation to the planet's orbit cannot be described in terms of a center belonging to its own totality. The sufficiency and closure attributed to the circle could not allow this merely computational device to be accepted as anything but a fiction.

Some historians condemn Copernicus for objecting in principle to the equant. Such a philosophical prejudice, they argue, hardly belongs in the domain of mathematical astronomy.[49] It is true that the reintroduction of the equant helped put Kepler on the road to the discovery of elliptical orbits.[50] Nonetheless, it may seem "illegitimate to impose on Copernicus a dogmatic conception of what belongs and does not belong to mathematical astronomy, and to require that he not be influenced by any external elements."[51] Is the importance attached to circularity an *external element*? Doesn't the decision to distinguish between interior and exterior itself belong to the *interior* of a particular conception of astronomy and its history? For Copernicus, in any case, the preoccupation with circular perfection does not come from somewhere else, from a kind of discourse foreign to the work of the astronomer. Since the universe is the work of a perfect artist, the astronomer cannot help but be concerned with the perfection of the qualities that he sees there. We find ourselves confronting an effect of the presuppositions at work within Copernicus' discourse.

CHAPTER FIVE

# Metaphor of the Center

## A Controversial Passage

In Book One, chapter 10 of *De revolutionibus*, after he has finished laying out the symmetrical order of the universe, Copernicus turns to lyrical praise of the sun's centrality in the universe he proposes:

> At rest, however, in the middle of everything is the sun. For in this most beautiful temple, who would place this lamp in another or better position than that from which it can light up the whole thing at the same time? For, the sun is not inappropriately called by some people the lantern of the universe, its mind by others, and its ruler by still others. The Thrice Greatest [Hermes Trismegistus] labels it a visible god, and Sophocles' Electra, the all-seeing. Thus indeed, as though seated on a royal throne, the sun governs the family of planets revolving around it.[1] (OR, p. 22)

This is an often cited passage. Koyré goes so far as to say that it identifies for us "the deepest motivation of Copernican thought."[2] Frances Yates calls particular attention to the reference to Hermes Trismegistus and notes that "it is, in short, in the atmosphere of

the religion of the world that the Copernican revolution is introduced."[3] For Yates, the passage becomes an important argument
affirming the role of the hermetic tradition in the birth of modern science.[4]

Without a doubt, Edward Rosen is right to judge that one
should not isolate the allusion to Hermes Trismegistus from its
context. It appears, after all, in the midst of references to other
figures – to Plato, Cicero, Sophocles, and Pliny – not all of whom
can be classified as writing in the hermetic tradition.[5] It is true
moreover, as we shall see, that the geocentric cosmos did not in
any way obstruct the development of a significant solar symbolism. The passage to Copernicanism was not absolutely necessary
for the development of that theme.

Should we go farther and agree with Jean Bernhardt that the
entire passage is "suspect of being purely literary"? Given that
the Copernican sun plays no dynamic role in the motion of the
planets, Bernhardt concludes that the metaphors likening the
sun to a "ruler" or "governor" are exaggerated and unsatisfactory. Throughout the passage, Copernicus seeks solely "to provide external support for his arguments and to make the position that he assigns the sun appear less of an innovation than it
really is."[6]

An expression like "suspect of being purely literary" is of
course itself suspect. It would be more neutral to speak of a stylistic marking, without introducing by choice of terms the idea
of an inverse proportionality between the relevance of the information conveyed and the visibility that a style guarantees for the
passage. On the other hand, although the sun does not effectively
participate in a dynamic relation with the motion of the planets,[7]
it is nonetheless true that it "governs," that it is the "ruler," the
center of the "symmetry" according to which the relationship
between distance and time is organized. A point of reference for

the cosmicality of the cosmos, it introduces into homogeneous space, as we have seen, a *perspectiva superior*.

The decision to treat the passage as an "exterior" element obviously arises from a modern prejudice concerning the separations among "science," "literature," "philosophy," and so on. Elsewhere, Copernicus wrote that astronomy leads to the "contemplation of the highest good" (OR, p. 7). His meditation on the meaning of the sun's centrality invites precisely such an act of contemplation, and that explains the passage's lyricism and proliferation of devices such as the rhetorical question, enumeration, asyndeton, metaphor, and comparison. Rather than consider such passages as "literary," why not recognize in their coherence and insistence a specific constituent element in the comprehensiveness and unity of the Copernican enterprise.

In any case, it is undeniable that this stylistic marking influences how the passage is read, and that it has acted as a stimulus on readers as different as Brahe, Kepler, and Mersenne.[8] Thus we cannot ignore it. The issue for us is to examine how the passage makes sense with respect to the intertextual unit that it mobilizes, how by appropriating traditional metaphors it establishes a new symbolic perspective on the organization of the universe.

## From the Midpoint to the Center

### Sun Worship

Copernicus' images and sources must be understood in the context of a larger movement that from the fifteenth century on, and especially in Italy, gave rise to veritable sun worship. Eugenio Garin has emphasized that the development of a solar myth is linked to the reawakening of Platonism and return to the *prisci philosophi* [ancient philosophers], and that it became widespread in non-Aristotelian writing.[9] Ficino played a dominant role, not only translating Hermes Trismegistus, but also writing long celebrations of the sun (*De sole*) and light (*De lumine*), and lavishing praise on the solar star in his other works. Thus we read in *Theologica platonica*:

> Who is God? A spiritual circle whose center is everywhere and circumference nowhere. But if this divine center possesses in some part of the universe an imaginary or visible seat for his operation, it is rather from the middle that he reigns, like the king in the middle of his city, the heart in the middle of the body, the sun in the middle of the planets. He placed his tabernacle in the sun and in the third essence, the median essence.[10]

Numerous authors took up this solar theme. Pico della Mirandola delivered a long eulogistic commentary on the Biblical verse *In sole posuit tabernaculum suum* [He has set his tabernacle in the sun], emphasizing that the sun occupies the "middle of the entire heavens."[11] Celebration of the sun persisted through the sixteenth century and spread into all genres.[12] The school notes of the young Galileo, whose Pisan professor adhered to the geocentric concept of the universe, show traces of the theme:

> The sun is the king and like the heart of all the planets, from which it follows that it must be placed at the center, for the king is in the middle of his people and the heart in the middle of the animal so that they can provide equitably for the needs of the people or the members of the body.[13]

The theme of the sun, which as governor or king of the stars is God's visible representative, also exercised a deep fascination within the framework of a geocentric cosmos. Among other things outside the literary and philosophical domains, the spread of solar monstrances shows evidence of the impact of the theme at the level of a general mental set unlinked to a transformation of the image of the universe.[14] At first consideration, no symbolic "necessity" requires a shift to a heliocentric cosmos. It is nonetheless fitting to note that Copernicus' universe, while drawing support from an existing tradition that developed within another cosmological framework, profoundly transformed what it borrowed and realized potential meanings that hitherto had been blocked. The *De revolutionibus* does not *reflect* a solar myth, but *works* it, remakes and perfects it, thanks to its new cosmological premises.

## Copernican Symbolism

In the geocentric universe, approbation for the sun's position flows from a spatial symbolism that gives priority in the first instance not to the center, but to the high and the low. This position, moreover, does not represent an absolute center, but a midpoint. Finally, it is the position of a celestial body in motion, not at rest.

Composed of two superimposed parts, the sublunary region and the supralunary region, the geocentric universe is hierarchical, arranged in tiers. Aristotle is emphatic: "The view, urged by some that there is no up and no down in the heaven, is absurd." According to this conception, the upper "extremity" is "in nature primary."[15] Although the universe is spherical, it is organized along a *vertical axis* that divides it into qualitatively different regions. In such a context, the notion of "center" lends itself to pluralistic usage:

• In addition to the center of the sphere, it is possible to distinguish several centers or midpoints according to the divisions along the vertical axis between different regions.

• Since a qualitative distinction is superimposed on the distinction among places, the terms "center" and "middle" lend themselves to uses motivated by such qualitative distinctions.

Quantitatively, the geometric center of the sphere of the universe corresponds to the earth's center, that is the lowest place,
the least dignified, the place where Christianity locates Hell.[16]
But two other centers or midpoints are to be found along the rectilinear axis that links the geometric center to the circumference:
the earth's surface, situated halfway between the center of the
universe and the center of the moon, is the middle of the "physical" universe, whereas the sun occupies the middle of the "metaphysical" universe, around which the planets are ranged as around
a king. Of these three centers, the sun is incontestably preeminent, because it is the highest. But a conception that grants preeminence to the highest location insures that solar symbolism
will be subjected to certain reservations and difficulties. In *De
sole*, Ficino emphasized the opposition (and subordination) of the
celestial sun to the divine *super*-celestial principle.[17] The frontispiece of Gafurio's *Practica musice* (Figure 5.1) clearly illustrates
the subordination of circular thinking to axial thinking. To be
sure, there are concentric spheres, but they are dominated by the
vertical figure of a three-headed serpent connecting the highest
heaven to the center of the earth. Apollo the sun god, moreover,
is represented twice. He appears in the emblem accompanying
the sphere of the sun, but also and more prominently *above* the
highest circle (directing the chorus of muses and celestial harmony).[18] The domination of the vertical over the circular and the
doubling of Apollo with a qualitative difference show just how
the universe is imagined primarily from bottom to top.

A famous work by Raphael, the *Disputa del sacramento* (1509),
containing a solar monstrance (Figure 5.2), supports a similar
conclusion. From our perspective, two findings are important.

On the one hand, a preparatory study preserved at the British Museum shows that Raphael had initially envisaged a chalice under the Host. The substitution of the solar monstrance for these objects associated with the Mass is another sign of the new taste. On the other hand, the composition of the whole remains firmly committed to the traditional cosmos. It even recalls the plate in *Practica musice* showing superimposed half-circles; an appropriate figure (monstrance, Holy Spirit, Christ, the Father) is associated with the middle of each circle. Here also, the solar symbolism is subjected to a spatial symbolism based on the opposition between high and low.

A vertical hierarchy of matter and being makes it possible, moreover, to combine the quantitative notion of "center" with the idea of "intersection" between entities with distinct qualities. Thus the human soul can also become a middle, both quantitative and qualitative. On the one hand, it corresponds to the third of the five orders of being and thereby constitutes, for Ficino, the arithmetical center in the hierarchy of creation.[19] On the other hand, the soul participates in two different natures, one celestial and the other terrestrial: it represents a qualitative mediation, a midpoint where the inferior meets the superior, an intersection, "the middle term of all things, the link and juncture of the Universe."[20]

In terms of symbolic activity in general, the proliferation of "centers" in the geocentric cosmos is not an exceptional phenomenon. According to Mircea Eliade, the symbolism of the center usually dispenses with geometric implications.[21] Ficino, moreover, is explicit on this point; he does not seek mathematical precision, because such calculations are "often less useful than they assuredly are difficult."[22]

If the symbolism is mathematically inaccurate, Ficino is *conscious* of that inaccuracy. And so are others, who echo Aristotle's

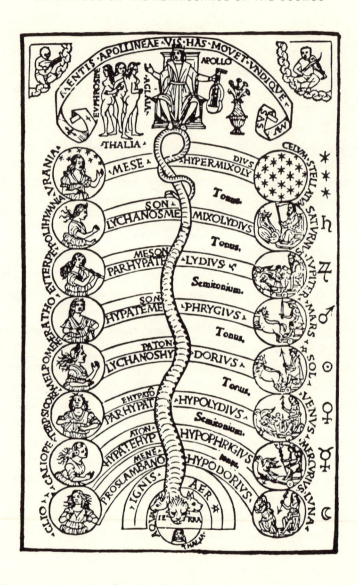

FIGURE 5.1. Frontispiece from Franchino Gafurio, *Practica musice* (Milan, 1496).

FIGURE 5.2. Raphael's *La Disputa del sacramento* (Vatican, Sala della Segnatura, 1509). The interference of verticality and circularity in the organization of space.

discussion of this point.[23] In his commentary on the *Pimandros* (where he grants a strictly instrumental value to the Copernican system), Foix de Candale accepts the imperfection in a geocentric cosmos of the solar metaphor for God. The image is based on a symbolism that is "philosophical" not "geometric." To be the latter, it would have to consider "quantity and motion."[24] To take on a geometric meaning, the sun "would have to be at the center," and occupy "the geometric center" of the universe, not the middle of the spheres: the Copernican system would be required. Moreover, in a geocentric cosmos, the sun moves. Although this motion is more perfect than the motion of the other planets (it has no epicycle), it is nonetheless opposed to the divine quality of *repose*.[25] Once again, the Copernican universe is more in accord with the symbolism than is the geocentric universe. The sun is *at rest*, stationary, *as though seated on a royal throne* (OR, p. 22).

Even though the symbolism of the sun can dispense with geometric precision, we must still acknowledge that such precision augments its suitability. Eliminating the distinction between high and low, basing its symbolic meaning on the relation between a center at rest and its periphery, the Copernican cosmos realizes more perfectly the conditions that allow solar symbolism to flourish. We cannot say that Copernicus simply picked up a traditional theme. He perfected it, overcoming acknowledged imperfections. Doesn't that happy outcome motivate a lyrical outburst?

*Parallels*

The panegyric of the sun enthroned at the center of the universe echoes a debate concerning the symbolism of another sacred place, the church. Francesco di Giorgio and others engaged in a controversy over the appropriate location for the church altar: should it be at the center of the church or on the periphery?

Those in favor of placing it on the periphery interpreted the distance from the entrance to the altar as a symbol of the infinite distance separating man from God: "Many say that to show that God surpasses us infinitely in nobility and perfection, the tabernacle should be as far as possible from the main entrance, and this place is only found near the part of the circumference opposite the entrance."[26]

The altar should be as far away as possible, *like* the Empyrean heaven in the geocentric cosmos. For Ptolemy, the original principle of Creation "is absolutely apart from sensible things." Placed at a distance that makes it inaccessible to human thought, it coincides with "the first cause of the first movement of the universe," located "high above, somewhere near the loftiest things of the universe."[27] The pertinent symbolic opposition is not between the center and the periphery but between the near and the far, the low and the high. The noblest place is the highest or most distant place.

Those, on the other hand, who preached that the altar should be placed at the center of the church, used the same argument as Copernicus. The right place in the church for God's image is at the center of the family of the faithful, *as* it is right that his cosmic symbol should be at the center of the planetary family:

just as God is present in every place and every creature, and as He is, for this reason, the preserver to whom all creatures are bound, it seems appropriate that the sacrament or the image should be at the center of the temple, since all radii from the circumference converge on the center and are linked to it. On the other hand, just as Christ teaches us, there where several are joined in his Name, He is in the middle [of all]. It is right therefore that the sacrament and image should be at the center of men joined to praise Him in the temple. And temple. And since at the circumference there are numerous places of common value and dignity, and since the center is unique and absolute, raised above all others, this center seems suitable [for God], by analogy with Him who alone truly is and with respect to whom all other things are merely shadows.[28]

Giorgio was doubtless inspired by the same Platonic and Pythagorean sources as Copernicus. It is impossible not be be struck by the similarity. Wittkower notes that most churches with a central altar were constructed between 1490 and 1530 (corresponding, interestingly, to Copernicus' stay in Italy), after Brunelleschi set an example, as early as 1434, with his plan for Santa Maria degli Angeli in Florence.[29] Thus the application to architecture preceded the application to cosmology.

From church architecture, the exaltation of the center spread to town planning, the planning of utopian towns in particular. Robert Klein emphasizes this relationship, and its larger ties with the Renaissance: "The radial town is no more than the urban variation on the central plan of religious architecture. It is a Renaissance form, and valuable as such." At the same time, he finds in these towns a cosmic model: "The utopian town is easily a reduction of the cosmos." Of special interest from our point of view, he adds: "But the characteristic obsession of the utopians was...the image of the sun."[30] Radial town, solar town.... To be sure, the complexity and diversity of utopian motivations (not only religious, but also political, economic, social) must put us on guard against abusive generalizations and assimilations. It is nonetheless true that a good number of these towns have at their center a place (religious and/or profane) representing the spiritual or intellectual "sun" of the community.[31]

Churches and towns: these forms favoring the center are not

unimportant in an intertextual reading of the Copernican enterprise. The universe is a *temple* (OR, p. 22); it is *sacred*, perme-
ated with symbolism. Artists tested and developed the symbolic,
sacred potentialities of the form used by Copernicus. This form
appeared in a context where it made sense, or rather, where it
made more and more Sense. Artistic treatment added symbolic
prestige to astronomical "grammaticality."

The predilection for the center and its periphery is expressed
not only in the imagination of concrete places like a church or
town, but even more in the work of Erasmus and the evangelical
conception of the relation of the faithful to God. The idea of a
concentric community, with the Gospels at the center as the solar
principle, replaces the conception of a Church organized hier-
archically from high to low. A faith where hierarchical interces-
sion is contested in favor of direct communication between a
center and a homogeneous periphery rises up against a multi-
tiered faith where communication passes between levels and
interposing authorities, and where the simple believer is further from God than the theologian – "thus what is known by
the learned is believed by the simple" [*Quod ergo a docte scitur, a
simplice creditur*][32]:

There are few learned men, but no one has ever been forbid-
den to be a Christian, to have faith, I even dare say, to be
a theologian.[33]

We...define so many things that without harm to our
salvation could be either unknown or left in doubt. Is the
man who does not know how to explain what distinguishes
the Father from the Son, and both from the Holy Spirit,
to be denied participation in the Father, the Son, and the
Holy Spirit?[34]

## The Language of Mystics

Like the solar metaphor, the quest for the center appeared regularly in mystical discourse. Thus, according to Theresa of Avila's biographer, the entire narrative of the *Moradas* grew out of her vision of a beautiful *crystal globe* (crystal sphere: figure of the universe) at the *center* of which dwells the *King of heaven*: "God showed her a very beautiful crystal globe, made like a castle with seven dwellings, and in the seventh, which was at the center, dwelt the King of Heaven in great splendor, illuminating and embellishing all those dwellings...."[35] A globe in the form of a castle: it is significant that in the text of the *Moradas*, the *globe* disappears. The center becomes uniquely the center of a *castle*. For of course Theresa of Avila was not concerned with the transpositions involved in the Copernican universe. And from the geocentric point of view, the cosmos does not lend itself to the metaphor of the search for a radiant center. The initial unrationalized vision is Copernican. The text, allegorized and recoded, is no longer Copernican.[36]

The quest for the center and the solar metaphor are necessarily dissociated in a discourse that refers to the geocentric universe. Thus, St. John of the Cross does not mention the sun in his long commentary on his verse: "In the deepest center of my soul...." St. John's gloss does not appeal to the sun or to Heaven; quite to the contrary, he refers to the Aristotelian physics of a sublunary, low world. The movement of the soul is comparable to the natural motion of a stone toward the center of the earth. Paradoxically, discourse on the most elevated experience must rely on comparison to the lowest:

> Accordingly, we assert that when a rock is in the earth, it is after a fashion in its center, even though it is not in its deepest center, for it is within the sphere of its center, activity, and

movement; yet we do not assert that it has reached its deepest center, which is the middle of the earth. Thus it always possesses the power, strength, and inclination to go deeper and reach the ultimate and deepest center; and this it would do if the hindrance were removed. When once it arrives and has no longer any power or inclination toward further movement, we declare that it is in its deepest center.

The soul's center is God. When it has reached God with all the capacity of its being and the strength of its operation and inclination, it will have attained to its final and deepest center in God....[37]

Pierre de Bérulle explicitly uses the Copernican metaphor in a passage worth citing in its entirety:

The AEGYPTIANS worshiped the Sun, and through excess called it the visible Son of the invisible God. But Jesus is the true Sun who looks upon us with the rays of His light, who blesses us with His countenance, who rules us by His motions: Sun that we must always keep our sight fixed upon and always worship. Jesus is truly the one Son of God; and no other thing, neither the Sun nor any other created thing, whether in Heaven or on Earth, can compare to Him. Jesus is the only Son, and the visible Son of the invisible Father, as we will say elsewhere. Let us say now that He is the Sun, not of the Aegyptians, who were deceived by their fables, but of Christians instructed in the school of truth and born of the light of that Sun, which is the Sun of the supernatural world, and a Sun that wanted to depict and represent Himself in this Sun, which is only His shadow and figure. For the Sun is the image of God, the Father of Nature, the universal Principle of life. And Jesus is the true and living Image of the Eternal Father.

He is His image, and sacred in His divine person, and even in His humanity He is united to the Divinity. He is the Author of the world, the Father of human nature, both by His Power in producing it, and by His Love in redeeming it. He is the Source of grace, and the Principle of the true life, in Earth and in Heaven, in Time and in Eternity, in men and in Angels, in Grace and in Glory. The Sun was formed in the middle of the days devoted to the Creation of the World, and it was placed in the middle of Creatures, above some and below others, to bring light to all. And Jesus, Splendor of the Father, appeared in the World, and came into the World through grace, to the middle of Time, at the end of the old Law and the beginning of the new Law, illuminating with the light of His grace the Fathers who had come before, and those who followed, both according to Scripture, as the luminous stars in the light of the Sun, in the middle of whom He rose and appeared to the World. And like created light, which had subsisted from the first day of the world, [He] was united to the body of the Sun on the fourth day, to be in him and through him a body and a principle of light on Earth and in Heaven. Thus the eternal light, light not created but uncreated, the inherent light of the Divinity, is in the fourth millennium united and incorporated in the humanity of Jesus to make in Him and through Him a Body and a Principle of life, of grace, of glory, and of light for all Eternity. One of the most famous Astronomers of Antiquity was so much in love with the principal object of his science, which was the Sun, that he wanted to be able to see and contemplate it from close up, and was burned and consumed by looking directly at it. Jesus is the object of the science of salvation and the science of Christians. The Doctor and Apostle of the World publishes on high that his science is knowing Jesus. Are not Christians

seized with love and desire to see and contemplate this principal object of their Belief, of their Science, and of their Religion? Will they not have more affection for the Sun of their
Souls than this Philosopher had for the Earth's Sun; Son exposed to sight and for the benefit of both men and beasts? And
do they not ardently desire to approach this Sun of Justice to
be not consumed but ignited with the fire of Love and Char-
ity while looking at Him? *An excellent mind of this century
wanted to maintain that the Sun is at the center of the World and
not the Earth; that it is immobile and that the Earth in proportion to its circular form moves in the sight of the Sun: by this contrary position satisfying all the appearances that oblige our senses
to believe that the Sun is in continual motion around the Earth.
This new opinion, little followed in the science of the Stars, is useful, and must be followed in the science of salvation. For Jesus is
the immobile Sun in His greatness, and the mover of all things.
Jesus is like His Father, and seated to His right, He is immobile
like Him, and gives motion to all things. Jesus is the true Center of
the World, and the World must be in continual motion toward Him.
Jesus is the Sun of our Souls, which receive all grace, light, and
influence from Him. And the Earth of our Hearts must be in continual motion toward Him, to receive in all its powers and parts
the favorable countenance, and the benign influence of this great
Star. And so let us direct the motions and affections of our Souls
toward Jesus....*[38]

Henri Brémond made famous the last part of this passage (cited
in italics) in his analysis of Bérulle's *theocentrism*.[39] The whole passage elicits the following commentary:

1. It begins with an allusion to "Aegyptians," that is to worship of the sun as visible god, midpoint in heaven, thus recalling
the revelations of Hermes Trismegistus, and according to Jean

Dagens probably influenced by François de Foix's commentary on the *Pimandros*.[40] The "Egyptian" theme of the spatial midpoint is associated with the cabalistic theme of the temporal midpoint. The sun was created on the fourth day of Genesis, and Christ came to earth in the fourth millennium, at the midpoint in the total duration of the history of the world. This parallel could have been inspired by Pico della Mirandola's *Heptaplus*.[41] Clearly Bérulle is heir to the Renaissance humanism of Ficino, Pico, Foix de Candale, to their reverence for the *prisci philosophi*, and to their syncretism.

2. The last section is Copernican. The focus shifts from the midpoint to the center. Bérulle's center exercises an attractive force on souls. Jesus is the sun, "the mover of all things." He "gives motion to all things," and "the World must be in continual motion towards Him." Dynamic center, motive force, which for this reason is no longer completely Copernican, but rather Keplerian. The sun of the *De revolutionibus* is effectively a purely geometrical point of reference and center of luminous radiation: "in the celestial *mechanics* of Copernicus, the Sun plays a very unobtrusive part. It is so unobtrusive that we might say that it plays no part whatsoever."[42]

3. The Copernican system is introduced with reservations: "new opinion, little followed in the science of the Stars." Bérulle distances and disengages himself from the question of the validity of the heliocentric model. Such caution notwithstanding, the same *opinion* "is useful, and must be followed in the science of salvation." Even though the referential foundation of the comparison may be only imaginary, it still furnishes a pretext for an auspicious transfer of meaning. The cosmological reference is stripped of reality; it is an "opinion," an improbable intellectual view. Bérulle wanted to make use of a metaphor of purely human invention, uncertainly anchored, rather than a firm interpreta-

tion of the true order of things. The rhetoric of figurative speech is detached from the fixed order of things.

4. Concerning the Copernican theme, Clémence Ramnoux notes, "thus was sketched a harmonious correspondence between infant science and classical religion. But the two separated...."[43] In effect, just as Bérulle's ideas began to spread, the Counter Reformation attacked not only the Copernican thesis, but also the diverse manifestations of a symbolism favoring relations between the center and the periphery. The Council of Trent reaffirmed the principle of the *vertical* and hierarchical organization of the religious community. Carolo Borromeo condemned churches with central altars, which he considered pagan.[44] In his *Almagestum novum*, Riccioli took care to assign limited significance to the relation between center and periphery. In the *supernatural* or celestial order, the place of greatest dignity is not the most central but the *highest*.[45] The mystical poetry of a Le Moyne, for example, returned to the orthodoxy of a geocentric universe and traditional physics:

> The spirit was raised to the circumference,
> And the body remained at the dispirited center.[46]

### Formistic Severity

In spite of everything that has been said about the emphasis on circularity, the vertical axis was far from disappearing from the Copernican universe. Special consideration was given to the relation between a center and its periphery, in terms both of the conception of the universe and how that conception was interpreted. The very fact that the universe continued to be tied to a transcendent meaning was founded on a vertical interpretive schema. The centrality of a stationary sun was conceived not only in horizontal terms, linking phenomena at the same level of reality in a syntagmatic chain, but also in vertical or paradigmatic terms, pointing to a higher level for which the phenomena are concrete manifestations.

Not only does this apply to the explanation of the centrality and immobility of the sun; it also motivates the synecdoches we have encountered, synecdoches that characterize Copernican physics and the elimination of the equant. In this physics, spherical perfection and the conservation of integrity are final causes. The equant is resented as a disturbance of the perfect and circumscribed relation between a center and its circumference. The universe is conceived in terms of teleological and formist perfection. Anything that puts that perfection at risk must be excluded. The task of astronomy and cosmogony is to *reduce the divergence* between phenomena and circular perfection in the work of the supreme artist.

Just as the theme that the universe is a "symmetrical" work of art confirmed and corroborated its symbolic character, so did tropological effects linked to the formist hypothesis of circularity. The universe is the *analogon* of a transcendent Meaning, which is unrepresentable in itself. The entire thrust of the Copernican construct is to improve the adequacy of the analogical model to its meaning:[47]

1. The first prerequisite for the proper functioning of an analogy is a precise description of what is being compared. The Copernican universe, which unites all phenomena in a single coherent system and strives for "certitude," is a better analogical model than the geocentric universe, which is less coherent and even appeared to some as a monstrous anti-model.

2. An internal structural requirement for any analogy concerns the clarity or precision with which the comparison is *applied*. In this respect as well, the Copernican universe, with its single geometric center, is superior to the other model, with its

multiple, ambiguous "centers" that are sometimes quantitative and sometimes qualitative.

3. The Copernican analogy is richer and denser, since it augments the number of predicates common to the source and object of the comparison. In particular, it adds the sun's immobility to previous motivations for making it a figure for divinity.

4. The elimination of the equant produces a more systematic analogy. For Copernicus, the velocity and orbit of a planet can be represented as a function of the same center. The suppression of an unsuitable center removes an ambiguity and thereby augments the perception that this analogical model is constituted of mutually constraining elements.

If the symbol appears more adequate for what it symbolizes in the Copernican universe, it is appropriate to note that the Copernican model represents a shift in the orientation of symbols. Geocentric analogies focus on concrete elements (the qualities of the sun: heat, light...) and only a few incomplete relationships (the sun, the "outer" and "inner" planets, but not the moon or global "symmetry"). The Copernican model, without totally ignoring the elements, increases emphasis on relationships ("symmetry," the elimination of the equant, the center as geometrical place). Since, moreover, the proposition that the earth moves directly contradicts sense perception and is justified on purely rational grounds, it is clear that Copernican symbolism presupposes a much higher degree of *abstraction* than geocentric symbolism. The transition to Kepler will accentuate this shift.

# PART TWO

# Kepler or the
# Mannerist Cosmos

It is impossible to exclude the astronomer
from the circle of philosophers who inquire into
the nature of things.

– Kepler

# CHAPTER SIX

# From Copernicus to Kepler

## Aspects of the Reaction to Copernicus

Toward the end of *De stella nova* (1606), Kepler enumerated at length the political, cultural, and other upheavals that had occurred during the past hundred and fifty years. The passage shows that he thought of his situation as an astronomer in the context of the *total* phenomenon that we call the Renaissance: the particular evolution of astronomy conforms to a general model. The passage begins with the widely used image of a man waking up after the long sleep of the Middle Ages. What follows describes first the decline of civilization since the end of antiquity, then – in detail – the accomplishments of the century and a half that had just passed. This account concludes by situating astronomy as part of a global movement: "Today a new theology and a new jurisprudence have taken shape, the followers of Paracelsus have renewed medicine, and the Copernicans astronomy" (GW, vol. 1, pp. 329–32).

If Copernicus "restored" an ancient astronomical theory and thereby took part in the work of the Renaissance, we must add that by Kepler's time this theory had been subjected to different evaluations and interpretations, and that it was far from having supplanted the Ptolemaic system. *De revolutionibus* introduced a higher degree of coherence, but it also posed new difficulties and provoked new questions. By advocating an alternative representation of the world, the first effect of Copernican theory was to create a dilemma.

The history of the reaction to Copernican theory is long and complex.[1] Our aim here is not to give a more or less complete picture, but to illustrate how the dominant "poetic" values that gave shape to Copernicus' text also shaped a system of responses, and thus make it possible for us to situate Kepler. It is important to remember that before Gilbert, who inspired Kepler, the formist explanation of motion had hardly been questioned. In general, we find a combination of irony and mechanism, or anagogy and organicism.

### Irony and Mechanism

At first, the truth claims of *De revolutionibus* were generally received with some reserve. There are several reasons for such a reaction. Other scientific transformations had to take place before physical evidence could corroborate the heliocentric theory. From the Aristotelian point of view, as Alexandre Koyré notes, Copernican physics is not without its weaknesses.[2] At the same time, the contradiction between heliocentrism and several Biblical passages inclined some individuals to prudence.[3] And even if Osiander's preface is not the simple negation it is habitually seen to be, but rather introduces the paradox I have already described, it nonetheless made the connection with truth ambiguous. Moreover, it moved in the direction of ironic skepticism, which is also a constant of the century.[4]

All these factors increased the possibility of an ironic reaction in the sense I have described. Beginning with his reading of the *Narratio prima* in 1541, Gemma Frisius declared that it mattered little to him whether the earth turns or is stationary. He felt unconcerned by the truth or falsity of the Copernican theory, and looked to it for no more than an improved tool for calculating time and distance, that is, to "save phenomena."[5] In the same spirit, Praetorius referred to "the liberty of astronomers to construct hypotheses."[6] François de Foix wrote, "Not that Copernicus wanted to verify that such a situation and arrangement of the universe is true; it is only a supposition that served his demonstrations."[7]

Even Mulerus, in charge of the third edition of De revolutionibus (1617), wrote that after twenty-five years of study, he still could not affirm the truth of the Copernican theory, while being no more satisfied with the geocentric model. His position is one of astronomical resignation:

> But what good is it to linger on hypotheses, which are nothing more than fictions with which men try in vain to discover the world system by analysis.... We must recognize the supreme wisdom of God the Creator and the weakness of our intelligence, which must regard with awe more than comprehension the world machine.[8]

Whether through prudence or conviction, from this point of view Copernican theory is unlinked to a world vision. An ironic reaction is combined with a mechanistic attitude. For if "symmetry" constitutes a thematic requirement that gave direction to the Copernican enterprise, we must admit that this requirement did not initially play a major role in the reactions to De revolutionibus.[9] Until at least 1570, readers were primarily interested in the treatment of certain specific problems in Books 2-6, without adopting the new model of the universe. The approach is clearly mechanistic, interested in assembling parts to describe specific phenomena rather than in the organic unity of the whole. This is especially true of those who adhered to what has been called "the Wittenberg interpretation." Reinhold, Peucer, and even Praetorius recognized the advantages of Copernicus' approach for determining the angular position of the planets or describing the movements of the moon, but they refused to concede that the earth moves.[10] Certainly they recognized the organic superiority of the Copernican model: Praetorius praised the "symmetry of all the orbs," but he nonetheless limited himself to small, discrete borrowings within a framework that remained geocentric.[11]

A typical but little-known example of the mechanistic reaction is furnished by Dodoens, famous for his *Herbarium*, but also author of a *Cosmographica isagoge*, published in Antwerp first in 1548, then in a new and corrected edition in 1584.[12] In the preface to the second edition, the author claimed that his work is preparation for those who intend to read Copernicus.[13] But he failed to breathe a word about heliocentrism. He was content, in fact, to take from Copernicus only his calculations regarding the size and proportional distances between the earth, the moon, and the sun.[14]

### Anagogy: Non-Copernican Organicism

Giordano Bruno and Tycho Brahe's reactions to the Copernican model diverge, but both are organistic. The terms in which they conceived their opposition was, at least in part, anagogical.

Both Copernicus and Bruno thought in terms of the principle of sufficient reason: Creation obeys a law of necessity that we must look for in God. But, whereas the *De revolutionibus* aimed at deducing divine perfection from the mathematical form of the cosmos, Bruno conceived of the universe as a necessarily infinite expression of the infinite power of its Creator: "Why would divine goodness, which can communicate itself and spread to infinity, want to be niggardly and hold itself to nothing, since any

finite thing is nothing with respect to the infinite?"[15] God's relation to the universe is not that of an Apollonian shaper, but of
power and force, which cannot be observed in terms of mathematical "symmetry." Bruno reproached Copernicus with "having
studied mathematics more than nature," with not having been
able to "free himself and others from so much vain research."[16]
The elimination of the equant, which constituted an important
achievement for mathematical astronomy, did not interest Bruno.
He did not retain the Copernican symbolism of the sun as geometric center of the universe. In a universe that is necessarily
infinite, the very idea of a center disappears, and the sun becomes
one star among many in continuous and homogeneous space. For
Bruno moreover, the sun, instead of being stationary, rotates
around its own center, for nothing can be at rest: in a universe
of infinite exchange, everything is in motion.[17] Finally, Bruno
rejected altogether the existence of Copernican "symmetry."
After having accepted it in the *De infinito*, he rejected it in *De
immenso* in favor of the principle of homogeneity. In an infinite
and homogeneous universe, it is impossible to differentiate the
planets according to the period of their revolutions.[18]

It is well known that Tycho Brahe proposed a modified geocentric model of the universe, according to which the moon and
sun orbit the earth, while the planets orbit the sun. His refutation of Copernican theory relied on classical arguments, such as
Aristotelian physics and Biblical authority, but also included a
detailed discussion of "symmetry." Just like Bruno, Brahe had first
been seduced by Copernican symmetry.[19] But subsequently the
Danish astronomer reproached one of his correspondents, Rothmann, a staunch Copernican, for "not giving enough weight" to
the implications of the increase in the volume of the universe
required by the Copernican system.

As already mentioned, Copernicus found himself obliged to

increase dramatically the dimensions of the universe in order to explain the absence of parallax in our observations of the stars. To satisfy a necessity inherent in the system, the volume of the universe abruptly increased by a factor of at least 400,000.[20] This increase applies to the distance separating Saturn, the outermost planet, from the fixed stars. By comparison, the dimensions of our planetary system remained practically unchanged. The universe suddenly contained a huge vacuum. It is this that Tycho Brahe found particularly shocking. In it, he could see "neither reason, nor utility"; it deprives Creation of regularity and order, "harmony or proportion." Not only were the stars at an immense distance, but the proportions were monstrous: to explain why stars of the third order were visible in spite of their distance, it became necessary to grant them a volume equal to at least the earth's annual orbit. How could we attribute to Creation "such large asymmetries," such "an ageometrical and asymmetrical foundation?" It is in this context that Brahe interposed the theme of the symmetry of the human body as articulated by Dürer. The passage has already been cited: it completed the refutation of the heliocentric universe.[21]

For Brahe, the Copernican universe was therefore synonymous with a new form of asymmetry. His argument would be employed often until far into the seventeenth century. Thus in *Vesta*, published in 1634, Froidmont asked what purpose there could be for such an expanse of emptiness in a universe created for man. To be visible, he noted, the smallest stars must be as large as Saturn's orbit. The enormity of these measurements deprives an eventual symmetry of any beauty: "For, between the parts whose mass is gigantic and arbitrary, there may be a proportionality, but not true beauty."[22] For Froidmont, symmetry is therefore not a sufficient condition for beauty.

Riccioli also referred to Brahe's reasoning and cited Dürer

once again on the subject of *symmetry*. He especially insisted on the contradiction between the ideal of a universe created *propter nos* and the distance of the fixed stars: "For who could believe that God had placed all the largest...stars at such a distance that we cannot observe [their] parallax?"[23]

### Anagogy: Copernicus' Defense
Altogether different is the attitude of those who adopted from the outset the new plan for the universe proposed by Copernicus. First among these was Rheticus, who praised "the remarkable symmetry and link between motion and the orbs, which are admirable and worthy of God the creator and of these heavenly bodies."[24] They also included Michael Maestlin, first the teacher of young Kepler, then his friend, who declared in his introduction to the *Narratio prima*, published as an appendix to Kepler's *Mysterium cosmographicum*:

> In effect, Copernicus' hypotheses enumerate, arrange, connect, and measure the order and size of all the orbs and spheres in such a fashion that it is impossible to change or transpose them in any manner without throwing the entire Universe into confusion. Yes, even any hesitation concerning location and linkage is excluded and held at a distance. In traditional hypotheses, on the other hand, the number of spheres is uncertain. For some count nine, others ten, others eleven.... The arrangement is doubtful: it is impossible to give any definite distance, except for the sun and moon, still less to demonstrate such distances. The debate on Venus, Mercury, and the Sun has not yet been laid to rest and never will be.[25] (GW, vol. 1, p. 83)

Also among this group, of course, was Kepler, who avowed

without circumlocution that his first concern was not any particular detail but the organization of the whole and the "philosophy" that accounted for it (GW, vol. 1, p. 9): "There were three things above all for which I tirelessly sought the causes: namely the number, dimensions, and motions of the orbs." Insisting on the anagogical meaning, he emphasized that "Copernicus has offered us a still unexhausted treasure of truly divine considerations concerning the marvelous arrangement of the universe" (GW, vol. 1, p. 16).

In response to Brahe's charge of "asymmetry" concerning the supposed absurdity of distances and size, the defenders of the Copernican system cited what they regarded as an equal absurdity, one that Copernicus himself had already identified: in the geocentric system, the stars are propelled by their sphere with a diurnal motion that supposes, in Maestlin's terms "a perpetual velocity that is unimaginable and beyond all credibility."[26]

In the second place, Tycho Brahe's computations were called into question. Maestlin remarked that the region of the fixed stars is *imperscrutabilis*, that it can be explored "with no instrument, with no contrivance" (GW, vol. 1, p. 437). This appeal to the lack of an adequate means of observation was pertinent. Later, Galileo would emphasize the errors committed by astronomers as long as they lacked the telescope; he proposed very different figures, diminishing the diameter of the stars and increasing still more their distance.[27]

Finally, in the framework of an anagogical justification, Maestlin argued that Brahe was guilty of human prejudice in judging the Creator's divine plan. Can one "prescribe laws to the omniscient Creator?" "Did mortal man assist the Spirit of God? Was man His councilor?" A kind of reasoning appeared that imposed limits, insofar as the region of the fixed stars was concerned, on the Copernican belief in a universe created *propter nos*. "Did

Tycho imagine that it was given to man to bind God's inscrutable wisdom to human senses?" (GW, vol. 1, p. 438). It was possible that the universe of fixed stars had been designed to remain "inscrutable," to defy human investigation. Kepler developed this idea more fully; moreover, he was not afraid to go much higher than Brahe in his estimates of the distance between the sun and the fixed stars. In *De stella nova*, he extended this distance to thirty-four million times the earth's radius (GW, vol. 1, p. 235). Only the unfolding of an anagogical reading of the Copernican universe can explain the acceptability of such a figure. For Kepler, as we will see, the partitioning of the universe implied by the vacuum between Saturn and the stars, like the impossibility of acquiring knowledge about the universe of fixed stars, is part of a significant design in God's Creation *propter nos*.

### Toward Mannerism

The specific problems posed by Copernican organicism concern the *decorum* of the representation. In rhetoric, the notion of *decorum* applies to the suitability of form and function.[28] If indeed the universe represents the perfection of the Creator and was created for man's use, the question arises whether the form Copernicus attributed to it is best adapted to assure its functions.

Copernicus conceived the form of the universe in terms of aesthetic requirements that are characteristic of the classical Renaissance: "symmetry" and the perfection of the circle. But satisfying the requirement for "symmetry" led to a dissociation between the real *form* of the (heliocentric) universe and our *image* of it. To be sure, astronomy had always been expected to correct errors introduced by subjective visual impressions.[29] But in this case, it produced a radical break between the tangible and the intelligible.

Ultimately, it is the dissociation between the form and image of the universe that Bruno radicalized, and in the very name of the *decorum* of the infinite: if, by its form, the universe represents the divine being, it can only do so by presenting the unrepresentable as such. It is therefore necessary for its form to be such that no "image" can contain it, and so to be infinite. God, moreover, cannot "appear" except in an unrepresentable form; organistic *decorum* cannot reside in the finite perfection of mathematical relationships. Bruno situates it in inexhaustible energy, with respect to which all things, including distance and time, are equal. It is also the requirement for *decorum* that underlies Brahe, Froidmont, and Riccioli's critiques: *What is it good for?* is the question they repeat. For them, the break between the tangible and the intelligible leads first of all to anxiety, as if it were able to reanimate the scholastic speculations about a God who takes pleasure in deceiving man.[30] How to maintain the thesis of a "world constructed for us" [*mundus propter nos conditus*]? Arising from the desire to prove the power of the human subject, Copernican theory finally makes the relation of subject to object more problematic than ever.

From this fact it follows that Copernicus was already close in outlook to Mannerism. This world, which appears "monstrous" in its geocentric form, and as a perfect "temple" in its heliocentric reality, corresponds to what Scipio Ammirato calls an art that "paints many monstrous things, but conceals within them many beautiful secrets."[31]

Copernicus not only made the subject–object relationship problematic by the content of his theory, but also by the very fact of asserting its truth. If a stationary sun introduces greater coherence, it also poses new difficulties, provokes new questions. Copernicus did not supply an immediately acceptable solution to a given crisis; instead he provoked a crisis. From that point on, astronomers were forced to make a problematic choice between

radically different models of the universe.[32] Even irony and mechanistic criticism became choices marked as such. The situation
grew still more complicated with Bruno, Brahe, Campanella,
Patrizi, etc. Mannerist artists were confronted with the same
problem, brought about by a similar evolution.[33] If Copernicus
made the relation between man and his perception of the world
order problematic, the theory of art simultaneously made the
relation between the artist and his representations problematic.
Verifying on the one hand the sometimes large gap between naive
perception and artistic representations, and refusing on the other
to attribute those differences to purely subjective causes, the theory of art finally posed the "question of the possibility of artistic production as such."[34] In Panofsky's terms – applicable almost
literally to the post-Copernican situation, "that which in the past
had seemed unquestionable was thoroughly problematical: the
relationship of the mind to reality as perceived by the senses."[35]
The theory of Mannerist art would attempt to surmount this
problematic situation by "an ever-increasing tendency toward the
speculative."[36] This is also the road taken by Kepler, starting with
his first work, the *Mysterium cosmographicum*. The epigraph clearly
states his anagogical aim:

> Why the world? What cause and what reason did God have
> for its creation? Whence the numbers? What rule governs
> such a large system? Why six circles, and what orbit accounts
> for their intervals? Why such a gap between Jupiter and
> Mars, which does not correspond to [the gap between] the
> first orbits? (GW, vol. 1, p. 4)

Posing such questions right at the outset gave Kepler's work
a highly speculative turn, for which Tycho Brahe would reproach
him,[37] but which he held in common with the Mannerists. Thus,

# CHAPTER SEVEN

# The *Semiosis* of the World

## Rejected Interpretations

### Play and Necessity

Kepler's texts are impregnated with a cosmogonic *semiosis*. They present the universe as a collection of signs that concretize ideal transcendencies. "Celestial physics" – the study of the natural causes of celestial motion meant to replace "Aristotle's theology or metaphysics"[1] – forms a substructure on which speculations about a world-sign that resembles what it signifies can be erected.

The signifying nature of this world arises from both play and necessity. On numerous occasions, Kepler affirmed that God played when He created the world. The object of the game was to model Creation on the Creator, thereby providing it with a sense: "For, in the act of making, God played, arranging therein the adorable image of the Trinity" (GW, vol. 2, p. 19). This metaphor of the game expresses God's original liberty. Once the decision to play the game had been made, it immediately became the most necessary of games. In terms of the principle of sufficient reason – whose importance for Kepler has often been emphasized – a perfect Creator necessarily realized a perfect world.[2] Now a perfect world could not be created except in the image of the only perfect model, God Himself: the necessarily perfect world must represent its Creator. The study of the physical world thus remains correlated to an anagogical disclosure. The bond between the world as effect and God as cause necessarily takes the form of a resemblance. The remaining issue is to determine the interpretative system that will make it possible to explain this bond.

*The Linguistic Interpretation*
Discussing the names of the signs of the zodiac Kepler noted that the partitioning of reality to which the names correspond does not reflect a natural division, but derives from distinctions based solely on man's need to organize his field of vision. It is not true that the zodiac is composed "naturally of twelve equal parts" (GW, vol. 1, p. 168). Thus over time, the pincers of Scorpio came to be seen as part of Libra: instability of form, where chance alone brings the mirages of resemblance into play. The names of the signs of the zodiac were invented, especially by farmers and navigators, according to one or another fortuitous similarity with terrestrial beings or objects (metaphor), or coincidence between a celestial configuration and a seasonal event (metonomy) (GW, vol. 1, pp. 173–74). It would be completely false to see in the figurative transfer of these names an index of a profound relationship, as several striking peculiarities make clear: "why isn't Aquarius the Water Bearer watery, and Taurus or Scorpio fiery? And why are Taurus the Bull and Capricorn the Ram feminine signs?" (GW, vol. 1, p. 178).[3]

Elsewhere Kepler criticized anagrammatic speculations, another means by which some attempted to make language appear as a fully motivated system of signs. To his mind, such attempts are often narcissistic. He confessed to having hoped himself, in his "idle youth," to find a secret meaning in his name.

But since he found nothing more satisfying than *serpens in akuleo* and *Seirenon Kapelos*, he gave up such preoccupations.[4]

Kepler did not look for an original meaning in Hebrew words that would reveal the essence of of the things they named, an essence miraculously preserved through the ages. Meaning in Hebrew is just as arbitrary as in other languages: examples like the words for *hand* and *finger* show that "in different languages, things are named according to different properties." An "infinity of [human] causes" can explain the similarities as well as the differences between languages (OO, vol. 7, p. 768).

If language does not maintain a secret connivance with things, the only valid approach consists in studying the laws that govern its concrete evolution and real workings. These things are of no interest to the Cabalists: "Neglecting the obvious causes of derivation and etymology, they affirm the divine origin of their language, persuaded that God, the creator of language, meant for the nature of things to be depicted by the transposition of letters and who knows what mysteries to be hidden in complete sentences" (GW, vol. 1, p. 175).

In summary, the only valid conception of language for Kepler is a philological conception, in the tradition of Valla, Erasmus, and Vivès. There is no point in searching for an original meaning that would reveal the key to things; the point is to investigate the concrete workings with respect to the human subject in his historical and mutable reality.

## The Numerological Interpretation

Kepler was just as critical of the union of things and numbers as of things and words. There were those who imagined that numbers could account for the partitioning of things, whereas in truth numbers take on value only when used to help analyze the things they are supposed to inform. It is particularly false to explain the generation of geometric figures from numbers: the existence of numbers arises from the necessity of distinguishing the qualities of geometric figures, without which there would be nothing to assign the numbers to:

> The nobility of numbers, admired especially by Pythagorean theology, which compares them to divine things, arises solely from geometry. For the angles of figures can be quantified not because the concept of their number precedes them; rather the concept of number arises from the fact that geometric realities contain multiplicity within them.... (GW, vol. 8, p. 4)

Kepler returned at length to numerology in his controversy with Robert Fludd, following Fludd's criticism of *Harmonices mundi*. Kepler charged his adversary with giving more weight to human traditions, qualified as "mystical" and "poetical," than to natural, verifiable facts:

> What he borrows from the Ancients, I derive from the Nature of things, and I constitute from its very foundations. He makes bad use of things that are confused (because of the diverse opinions of those who transmit the tradition) and incorrect. I proceed according to the natural order so that everything will be without error according to the laws of Nature and so as to avoid confusion.... (GW, vol. 6, p. 374)

To protect the correspondence between things and numbers, to remain faithful to an interpretation whose validity is presupposed, Fludd is even ready to falsify the representation of reality: "he divides the entire universe into three equal parts, knowing full well that they are not at all equal..." (GW, vol. 6, p. 375).

Repeatedly, Kepler contrasted his *diagrams* to Fludd's *hiero-*

*glyphics* or *pictures*. (Figures 7.1 and 7.2)[5] The diagrams are superior because they are free of any interpretive presuppositions: they represent a partitioning of reality based only on observation, unmixed with hermeneutics. Fludd's pictures, on the other hand, pass through no stage of pure analysis; the partitioning is always accompanied by interpretation.

Without a doubt, the methodological priority thus given to a partitioning of reality that attempts to be objective prepared the way for what we call "modern" science, for the independence and sufficiency of "horizontal" analysis, for the rejection of speculative goals. It is no less true, however, that Kepler's universe is still a signifying object in a "vertical" order. And his analysis is accompanied by a hermeneutics preoccupied with the organization of cosmological signs deployed along a symbolic horizon. Kepler went so far as to characterize this enterprise as a *geometric cabala*: "Naturally, I too play with symbols, and I have planned to write a little work, *Geometric Cabala*, which would treat the ideas of natural things in geometry..." (GW, vol. 16, p. 158).

If Kepler rejected several forms of "play" based on comparisons, that does not mean that he rejected the very category of resemblance. But in place of the chimerical scaffolding of the traditional cabala, which takes no account of the natural order of phenomena (Fludd) or the natural causes of phenomena (etymological and anagrammatic speculations), the geometric cabala substitutes interpretation subject to verifiable relations between things, as they can be established by observation and scientific explanation:

But I play in such a way that I do not forget that I am playing. For we can prove nothing by symbols, we can discover no secret in natural philosophy by geometric symbols, unless they agree with established facts. It must be proved with solid rea-

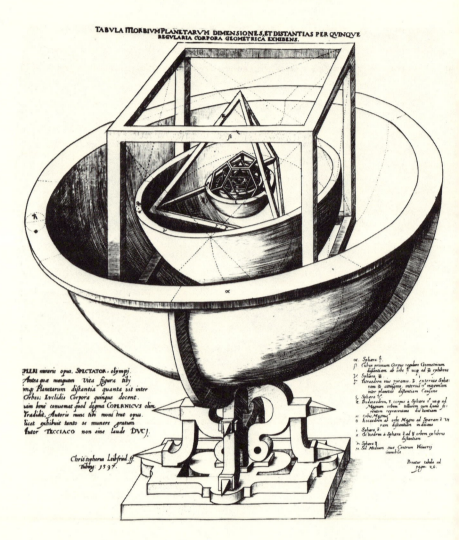

FIGURE 7.1. The nesting of polyhedrons: Kepler's first model of the universe, *Mysterium cosmographicum*, 2nd ed. (Frankfurt, 1621).

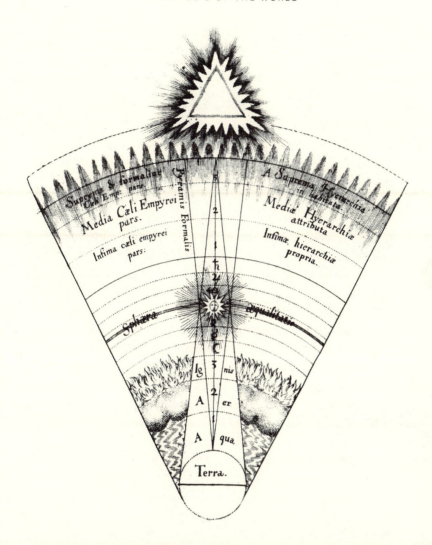

FIGURE 7.2. The pyramids of spirit and matter traversing the universe: Robert Fludd, *Utriusque comsi…metaphysica, physica atque technica historia* (Oppenheim, 1617), vol. 1. An example of the contrast between Kepler's "diagrams" and Fludd's "hieroglyphs."

sons not only that they are symbolic but that they describe the manner and causes of the interconnections of each element. (GW, vol. 16, p. 158)

The play of meaning in the world must be both possible and necessary. The preceding citation defines the criteria for its *possibility*: our reading of the world must respect the real disposition of things; we must not transform the world as we please, according to our desires or needs. The foundation for the *necessity* of this play of meaning is still to be determined. The object is to discover God's original interpretation, enacted in the Creation. Such a discovery would permit the cosmological astronomer to find the divine intention in a play without "play."

*Meaning and Communication*
If for Kepler the world continued to be "a book where God represented...his essence and his will concerning men" (GW, vol. 7, p. 25), the code in which the book was written is made up of geometrical ideas or archetypes: "where matter is, there is geometry" [*ubi materia, ibi geometria*] (GW, vol. 4, p. 15).

Geometric modeling is present throughout Creation, from the spherical form of the whole cosmos to the shape of leaves, fish scales, the fur of terrestrial animals (GW, vol. 6, p. 343), and even the hexagonal surface of snowflakes, to which Kepler devoted an entire small treatise.[6] Traces, "vestiges" (GW, vol. 1, p. 192) exist everywhere of geometry's role in the creation of the world.

To this intuition, Kepler attempted to add the necessity of deductive reasoning. In virtue of the principle of sufficient reason, the world must necessarily, as we have seen, represent its Creator. This representation must have possessed an original form, within God Himself, before being imposed on Creation: "the Creator of the world preconceived within Himself the idea of the world and the idea is prior to the thing..." (GW, vol. 1, p. 24). Before the world of material forms existed with the *qualities* that we can ascribe to it, the ideal form, coexisting with God as His own internal and immaterial reflection, must have been purely *quantitative* (GW, vol. 1, pp. 37–38). The world is necessarily informed by quantities and their relations: "quantities are the archetype of the world" (GW, vol. 15, p. 172). Since moreover, in the realm of quantities, numbers are secondary with respect to geometric extension,[7] the code in which the book of the world is written is a geometric code. God "eternally geometrizes" (GW, vol. 1, p. 26). The geometric archetypes are "coeternal" with Him (GW, vol. 6, p. 104), are in fact God Himself (GW, vol. 6, p. 223).

An inner conception precedes and governs the Creation of the material world. It determines the form of the world and from that point on constitutes Creation's final cause. Kepler's thinking recalls Thomas Aquinas' celebrated thesis: "For everything not engendered by chance, the end is necessarily a form. Now an agent would not act with a form in mind unless it contained within it a similarity to that form."[8] Beyond Aquinas, we can go back to Philo, to Plotinus, to Saint Augustine and the Christianization of the Platonic theory of ideas, which are no longer endowed with objective existence, but rather are contained within the divine *subjectivity*.[9] Kepler himself referred to Plato on this subject: as Creator, the Christian God permits himself to be conceptualized according to a modified Platonic scheme (GW, vol. 6, p. 265, and GW, vol. 8, p. 30). But it is worth recalling that in the second half of the sixteenth century art theoreticians insistently transposed this conception of the divine Creator to their realm. The concrete work of art, the *disegno esterno* [external design] is seen as dependent on the *disegno interno* [internal design], which is its inner and prior conception, its model and source, and which corresponds to the *a priori* apprehension of a and geometric archetypes are innate to him, such that observation does no more than elicit their recollection.[16]

God signifies and communicates to man through nature. Consequently, the practice of astronomy becomes a means of following the contemplative path that leads man to God. Insofar as this path leads man to the discovery of the plan of the cosmos as the image of its Creator, it can set up a circular relation that flows back along the signifying and communicative current from nature to God. Understanding the world is ultimately to find the Meaning and the Author in signs. The calculations with which the astronomer discovers his laws or supports his hypotheses contribute to making man more God-like. For the resemblance between man and God is as much a task as a fact. We construct the resemblance by reconstructing internally the laws that govern Creation. The result is an anamnesis: "Just as objects seen from the outside make us remember those we have already known, likewise the mathematics of the senses, if they are recognized, excite an intellectual mathematics, previously present to the inner [man], such that there actively shines within the soul what beforehand was hidden beneath the veil of potentiality" (GW, vol. 6, p. 226).

Once again, this network of communication and meaning is close to what is described by the theoreticians of Mannerism. For them also, the task of the recipient consists of working back from the exterior design to the interior design, to the *idea* that is signified. By reversing the direction of the flow of meaning, the recipient actualizes an idea that he had potentially within himself. Having discovered the interior design of the author, he becomes its active icon and signifies, as a man having (re)discovered his humanity, his identity to the other, in the sharing of the idea: "The recreative process finally implies a mutual exchange between creative *anima* and recreative *anima*, both expressing the same *idea* through the transformation of figurative material."[17]

Of course I do not mean to imply that Kepler's vision of the world was the source of the Mannerist conception of art. Nor the reverse. But the similarities are evidence of a common root in the problems arising from the Renaissance and common methods and tools for responding to those problems.

In art, the extensive speculations on the divine *disegno* and its human analogue can be explained in terms of the questions raised by Renaissance art. If a work is the result of a kind of synthesis, of an *electio* from what visible reality presents to us, what is the principle that guides the selection? What is the basis for discriminating between different works? What is the structure of the relation between reality, the artist, the work, and the spectator? What makes it possible to ascribe truth-value to a work, if it is true that it does not only reproduce nature? Together, these questions give priority to a *theoretical* approach to art, which is manifested in a systematic reelaboration of the theory of ideas, concerned with meaning as much as communication.[18]

Copernicus poses specific problems, but they lead to a comparable situation. How can we grant that our senses deceive us in our naive perception of the world, and that nonetheless the world was created *propter nos*? How can we conceptualize the relation between the Creator, His Creation, and man's perception and rationalization of Creation? And, if the aesthetic argument based on "symmetry" is really "the principal consideration," we are obliged to question God's plan, code, motivations, and intentions. Kepler took on all these question; he admitted their urgency. They led him to a systematic reflection on the divine idea of the world, on its material reality, and on how these relate to man. Like the Mannerists, he appealed to an entire philosophical tradition, which he attempted to integrate into his enterprise.

The Mannerists "recalled everything at their disposition in the realm of metaphysics."[19] Kepler turned to Pythagorean hermeticism, Plato, Proclus, Nicholas of Cusa, and so on, directing light from an entire speculative past on his specific object, thus producing a composite *manner* and an overdetermined conception of the world.

## The Geometric Code

### The Fundamental Opposition

For Kepler, the basic units of the geometric code are the curve and the straight line, representing respectively the Creator and his creatures.

> And God wanted quantity to exist before anything else so that that the curve and the straight line could be compared. Because they attached so much importance to the reciprocal relation of the curve and the straight line, and because they dared to compare the curve to God and the straight line to creatures, for this reason alone, I hold Nicholas of Cusa and some others as divine. To attempt to establish an equivalence between the Creator and his Creation, God and man, divine judgment and human judgment, is almost as vain as attempting to make the curved line equal to the straight line and the circle to the square.[20]

The opposition between the curve and the straight line, the circle and the square, the sphere and the cube, is a very widespread symbol of the opposition between the spiritual and the material. Plotinus, for example, distinguished spiritual circular motion from the "straight road characteristic of the body" (*Enneads*, II, 2). In his commentary on Euclid, Proclus had written: "Rectilinear figures are akin to sensibles, but the circle to intelligibles; for what is simple, uniform, and determinate accords with the nature of being, whereas to be diversified and to possess indefinitely more containing sides is a characteristic of sense objects."[21] Kepler himself pointed to Nicholas of Cusa as his principal source.[22] It is appropriate, however, to note two important differences, because they are at the origin of how the geometric theme specifically functions for Kepler.

In the first place, for Nicholas of Cusa geometric figures take on symbolic importance primarily as they are extended to infinity, where the curve and the straight line converge. His mode of thinking consists of a continuous passage from the finite to the infinite. This applies to both his reflections on geometric potentialities and how he imagines the actual extension of the universe. Kepler, on the other hand, posits the finiteness of the universe and develops his symbolic thinking in terms of a geometry of finite forms. Geometry serves a negative theology in one case, a positive cosmology in the other.

The other difference concerns the selection of the principal geometric figure. Kepler emphasizes it himself:

> For what Nicholas of Cusa attributes to the circle, and others to the entire globe, I only claim for the spherical surface. And I cannot believe that there exists any curve nobler and more perfect than the surface of a sphere. The globe cannot be reduced to the surface and includes a straight line which alone fills the interior. The circle exists only in a plane, for if the surface of the spherical globe is not bisected by the plane, there is no circle. (GW, vol. 1, p. 23)

The contrast between the curve and the straight line, combined with the selection of the spherical surface as perfect figure,

implies a geometric symbolism in Creation that is both structural and generative.

*Structural* Semiosis
On the one hand, the surface of a sphere was chosen as the shape
of the universe because of a strictly geometric property of the
sphere: "We are discussing the figure that encloses the universe
from without. Everything, consequently, must be inside this figure, nothing outside" (GW, vol. 7, p. 46). As Book Five of Pappus' *Mathematical Collection* demonstrates, the sphere possesses
the greatest "capacity" of any geometric form.

On the other hand, the spherical surface is most apt of all
geometric forms to become a symbol for God: "the archetype of
the universe is God Himself, which no figure resembles better
(to the degree that it is possible to speak of resemblance) than
does the sphere. For God is the being of beings, prior to all things,
unengendered, the simplest, the most perfect, unmoving, totally
self-sufficient and sufficient for all creatures, Creator and sustainer
of everything, one in essence, three in persons. In a rudimentary
mode, the surface of a sphere possesses...the same properties"
(GW, vol. 7, p. 47). Over a number of pages, Kepler describes in
detail the combination of qualities that make the spherical surface an appropriate symbol for God. Here we are interested only
in the most important of these.

If only the spherical *surface* — not to be confused with the
*volume* it delimits — composed of both curve and straight line,
can be accounted the most perfect geometric figure, this is primarily true because it is not derived from any other form; it is
self-engendered by the equal propagation of its center in all directions — propagating in such a way that "no vestige of it remains
in intervening space" (GW, vol. 7, p. 51). The regularity of its
surface, "everywhere equal to itself" and undivided, reflects the

simplicity of the center point; it displays this center, makes visible this point, which as such is "invisible" (GW, vol. 7, p. 51). It is the tangible revelation, the most appropriate representation of an origin that is itself invisible. Generated internally from the center it represents, the surface of the sphere is at once "cause and norm of perfection for all other geometric forms." For these forms, beauty is measured in terms of the degree to which they deviate from the simplicity of the spherical surface. "All imperfection resides in multiplicity, all beauty in simplicity" (GW, vol. 7, p. 49).

Moreover, such a structure is an exact image of the relation between the persons of the Trinity. "In the sphere, there are three elements: the Surface, the Center, and intervening Space. It is the same in the stationary universe: the Fixed stars, the Sun, and the Air or intervening ether. And it is the same with the Trinity: the Son, the Father, the Holy Spirit" (GW, vol. 13, p. 35). Kepler repeatedly returns to the spherical shape of the stationary universe as a structural representation of the Trinity[23]:

1. The center, image of the Father, is the origin of the form, the point that engenders the surface through an infinity of equal radii. Itself invisible, its location can be determined thanks to the convergence of radii leading from the surface to it.

2. The surface engendered by this "propagation" out from the center represents the Son. Each point on the surface reproduces the *same* point (the center point); it is the image of the center and its emanation.

3. The intervening space, which connects the center to the surface, the Father to the Son, represents the Holy Spirit. As that which both separates and connects, it is identical, moreover, to the other two, since it is constituted by a repetitive series (of points forming the radii) originating from the same point.

As J. Hubner notes, Kepler's language takes on Biblical over-

Creator. Since activity is only one of the aspects of divinity, it is appropriate that its representation should be enclosed within the representation of God's total and eternal being. And since activity defines God as Creator of the universe, the mobile universe involves the opposition between the curve and the straight line.

The sphere can be considered as the starting point for all other geometric figures. Kepler made a distinction between figures generated "mechanically" and those generated "metaphysically." Starting with the point, mechanical generation leads successively to lines, surfaces, and solids. Metaphysical generation is preoccupied with the correspondence between the geometric origin and the divine archetype. Since the spherical form corresponds to God's original representation of Himself, the "metaphysical generation" successively derives from this representation solids, surfaces, and lines. The latter are not "figures properly speaking, but the limits of figures" (GW, vol. 7, p. 47).

Generating figures from the sphere makes it possible to classify them according to a progression that corresponds to the order of God's intervention in the Creation of the cosmos. In metaphysical geometry, solid figures precede two-dimensional figures. The relation between them is like the relation of substances to qualities (GW, vol. 14, p. 640.) Or rather, solid figures are appropriate for structuring the creation of solids or bodies, whereas two-dimensional figures inform the properties of these bodies. The first are *somatopoetic*, the second *cosmopoetic* (GW, vol. 14, pp. 46). Consequently, the spacing of substances or bodies in the universe (the planets, that is) is based on the volumes of the regular polyhedrons – this is Kepler's thesis starting with *Mysterium cosmographicum* – whereas the harmonious relations (the qualities) of the constituent parts of the universe must be studied starting with two-dimensional figures (this is the subject of *Harmonices mundi*). In the selection of somatopoetic and cosmopoetic figures, the sphere and circle play a primordial role, which consists of eliminating the danger of an infinite proliferation of figures and of organizing into a reasoned hierarchy the ones that are retained.

It would be premature to follow in detail here Kepler's progress in his search for the deductive necessity of the universe. But at least I wanted to give an overall idea and show how metaphysical geometry, proceeding from the sphere to the polygons, can produce a model of the entire cosmos that is complete down to the smallest details. Nowhere else do we find an attempt to construct such a coherent and constraining system of the forms that give structure to the universe. If this endeavor is unique in astronomy and cosmology, it also seems to have been more or less foreign to the preoccupations of mathematicians. In Book Two of *Harmonices mundi*, Kepler revealed his discovery of two new regular polyhedrons, but this passage appears to have remained unnoticed until the nineteenth century.[26] As Gérard Simon writes, the questions raised by Kepler "are not mathematical, nor are they produced by the logic of mathematical development."[27] In fact, it appears that the principal references that permit us to free Kepler's geometric reflections from their isolation come from art theory, not only because of Kepler's sometimes critical but always sympathetic allusions to Dürer,[28] but more importantly because in the sixteenth century art theory was the primary field where regular polyhedrons and the harmonious proportions of polyhedrons were studied. It is worth recalling Piero della Francesca's *De quinque corporibus* (1492), Jamnitzer's *Perpectiva corporum regularum* (1563), Stoer's *Geometria et perspectiva* (1567), Barbaro's *Pratica della perspettiva* (1569), and Juan de Herrera's *Discurso de la figura cubica* (1589).[29]

A comparison with Dürer's work is particularly instructive. There is no lack of points in common. In particular, Dürer's treatment of the "congruence" of two-dimensional figures anticipates,

# CHAPTER EIGHT

# The Cosmographic Mystery

## The Mystery Elucidated

### From Symmetry to Eurythmy

In Copernicus' universe, the planetary spheres no longer touch; they are separated by the void of empty space.[1] Moreover, the distances between the planetary orbits seem to obey no rule and appear, in some cases, absolutely beyond all reason. Thus between Jupiter's perigee and Mars' apogee the distance is twice as great as from Mars to the Sun (GW, vol. 1, p. 48). Why such a gap? Why, moreover, are there six planets? Such were the principal questions broached by Kepler in his first work, *Mysterium cosmographicum*, published in March 1597. They were inspired by the desire to provide a better explanation for the plan of the universe, the design or *disegno* that God had proposed to himself in the Creation, and the excellence of which Copernicus had begun to unveil. Kepler's work begins with an eminently speculative essay, based completely on his belief in the principle of sufficient reason: "it is absolutely necessary that the work of such a perfect Creator should be of the greatest beauty" (GW, vol. 1, p. 23).

If the whole universe is an enormous sphere, the six planets, according to Copernican theory, revolve around the sun. Separated by five intervals, these planets may represent an inscription of the five regular polyhedrons. The objective of *Mysterium cosmographicum* is to prove that they do:

> It is in my intention, reader, to demonstrate in this little work that with the Creation of this mobile universe and the arrangement of the heavens, God the Great Creator had in mind these five regular bodies that have been so famous from Pythagoras and Plato to our days, and that he caused the number of the heavens, their proportions, and the system of their motions to conform to the motions of these bodies. (GW, vol. 1, p. 9)

The lyrical fervor of the work's epigraph conveys the author's enthusiasm for his discovery, as well as the importance that he attaches to it: "I die each day, and I confess it; but while my care keeps me hard at work on the roads of Olympus, my feet do not touch the Earth; in the presence of the Thundering divinity, I feed on nectar and ambrosia."[2]

The polyhedrons are nested within each other in the following manner:

> The Earth is a Circle that measures all the others. Circumscribe it with a Dodecahedron: the circle that contains it will be the circle of Mars. Circumscribe Mars with a Pyramid: the circle that contains it will be the circle of Jupiter. Circumscribe Jupiter with a Cube: the circle that contains it will be the circle of Saturn. Now inscribe an icosahedron in the Earth's circle: the circle that can be inscribed within it is the circle of Venus. Inscribe an octahedron in the circle of Venus: the circle that can be inscribed within it is the circle of Mercury. This is the reason for the number of the planets. (GW, vol. 1, p. 13)

It does not enter into the framework adopted here to discuss the specifically mathematical and astronomical problems posed by *Mysterium cosmographicum*: conformity with the numerical data, problems of a physical order, and so on.[3] On all these points, Kepler would vary, but he always remained attached to his first work, which he judged worthy of a new edition, albeit annotated, in 1621. He saw in *Mysterium cosmographicum* the source of the inspiration for his later works. In the preface to the second edition, he wrote: "almost all the astronomical books that I have published since can be linked to one or the other of the principal chapters of this little work, either as illustration or as improvement" (GW, vol. 8, p. 9).

The question that inspired *Mysterium cosmographicum* is derived from the Copernican preoccupation with "symmetry." Kepler's entire system of polyhedrons and nested circles is based on the position of the earth, and gives prominence to Copernicus' postulate that the earth constitutes a "common measure." In addition, whereas Copernicus invested celestial phenomena with greater coherence by relating the distances of the planets from the sun to the periods of their orbits, Kepler overdetermined the structure of the universe by supplementing this "symmetry" with a geometric explanation for the number of planets and the distances between them.

On the title page of the original edition, the words that stand out by the size of their characters, their spacing, and central placement, are not those that serve today to identify the book, but these: *de admirabili proportione orbium* [on the remarkable proportions of the spheres]. Above all, the subject of *Mysterium cosmographicum* is *proportion*. Maestlin emphasized this fact in the letter to the reader with which he prefaces Rheticus' *Narratio prima*, included as an appendix to Kepler's text: "by a very ingenious geometric discovery, our Kepler establishes the certain and finite number as well as the order of the orbs or spheres, and — which is more important — the exact size and the reciprocal proportion of planetary movements..." (GW, vol. 1, p. 82).

The notion of *proportion* was linked to the notion of *symmetry*, since, as we have seen, symmetry was defined as the "proper arrangement of proportions." The specific contribution of *Mysterium cosmographicum* to the conception of a universe dominated by proportionality can be better understood if we refer to the distinction between *symmetry* and *eurythmy* that Barbaro established in his celebrated commentary on Vitruvius:

> Symmetry is the beauty of Order, as Eurythmy is the beauty of Arrangement. It does not suffice to order the measurements one after the other, but it is necessary that these measurements should conform, that is that they should be well proportioned; and where there is proportion, nothing is superfluous.[4]

Copernicus had discovered the *symmetry* of the universe, that is its *order*, which is grounded in the agreement between distance from the center and period of revolution. But the precise reasons for the actual distances, the *arrangement* at such-and-such intervals, was not clear. To *symmetry*, the principle that regulates the linear ordering, *Mysterium cosmographicum* adds *eurythmy of arrangement*, the principle regulating both the magnitude of the intervals and the closure of the linear order. And so the work demonstrates that "nothing is superfluous" or without reason, neither the magnitude of the intervals, nor the number of planets. Everything in the universe can be explained by the plan that governs the whole. Copernican organicism overflows. Like the Mannerists, Kepler took up the aesthetic demands of the Renaissance, but multiplied their application to an increasingly overcoded universe.

*Earlier Attempts*

Before setting forth the hypothesis of the reciprocal embedding of polyhedrons and spheres – "orbs in bodies and bodies in orbs" (GW, vol. 1, p. 26) – Kepler had sought to explain the distances between the planets in other ways, which he reviews in the preface to *Mysterium cosmographicum*. In terms of the style of his cosmological speculation, these are representative of the investigations of aesthetic mathematics, based on a variety of sources and rich in intertextual connections.

In a first attempt, Kepler had examined whether one orb can be related to another as its double, triple, or quadruple, "or in a similar manner" (GW, vol. 1, p. 10). The inspiration for this idea probably came to him from the *Timaeus*, according to which the universal soul possesses such a structure according to two geometric progressions, one based on a factor of two (1, 2, 4, 8), the other on a factor of three (1, 3, 9, 27).[5] We should note that Renaissance artists based the proportions of their temples on the same source.[6]

When the results of this attempt turned out to be unsatisfactory, Kepler did not hesitate to modify the data, interposing invisible planets between Jupiter and Mars, Venus, and Mercury. This hypothesis may be have been inspired by the Pythagorean belief in an invisible "counter-Earth."[7] The new attempt was no more satisfactory than the first: "Even if by this means I obtained a particular proportion, the calculation had no rational basis; it was not based on the number of moving stars, with respect to the fixed stars... or with respect to the sun..." (GW, vol. 1, p. 10). Space could, in effect, be proportionally subdivided to infinity.

Another attempt, perhaps technically based on Regiomontanus' *De triangulis omnimodis*, relied on the sine and involved a
correlation between distance and period of revolution around the
sun.[8] Representing the propulsive forces of the planets by parallel lines extending to the arc of a circle, Kepler undertook to
examine their proportions. Once again, the attempt failed, this
time most notably because the value of the radius from the sun
to the fixed stars, which was to serve as the basis of the calculations, was an unknown (GW, vol. 1, pp. 10–11).

Then during a lecture, Kepler suddenly realized that it was
possible to inscribe an equilateral triangle between the spheres
of Saturn and Jupiter. The coincidence was striking: on the one
hand, Saturn and Jupiter are the *first* planets, starting from the
sphere of fixed stars; on the other, the equilateral triangle is the
*first* of the polygons, both in Euclid's *Elements* and the *Timaeus*.[9]
Immediately, Kepler attempted to inscribe a square in the circle
of Jupiter, a pentagon in the circle of Mars, etc. (GW, vol. 1,
pp. 11–12). Dietrich Mahnke draws a parallel between this attempt and a passage by Charles de Bovelles, but Bovelles started
with a square, bypassing the triangle.[10] On the other hand, there
is an analogy with both esoteric and aesthetic traditions concerning the human microcosm. Inscribing polygons within circles and
juxtaposing these figures with the human body was one way in
which the occult sciences represented the relation between spirit
and matter, a relation that Kepler himself represented with the
curve and straight line.[11] To this esoteric symbolism we can add
the Vitruvian tradition, which consists of inscribing the body
and its parts into regular geometric figures. The investigations of
Dürer, among others, would lead some of his Mannerist successors (such as Schön) to a "frantic exploitation of 'geometrism,'
with no purpose other than its own representation."[12] Did not
Kepler search for such a geometrism at the level of the macro-

cosm, making of the heavens a pure system of regular forms? In their theoretical writings, the Mannerists also superimposed aesthetic and esoteric sources. Thus Lomazzo, inspired by Cornelius Agrippa's writings on magic and occult science, described first how the regular polygons are derived from observation of the human body, then proceeded to the regular polyhedrons, making use of the aesthetics of Pacioli.[13] A similar superimposition of different traditions (Pythagorean, Platonic, Euclidian, artistic, etc.) applies to the Kepler of *Mysterium cosmographicum*, not only in those endeavors that turned out to be fruitless, but also when he finally settled on the regular polyhedrons.

*Pythagorean Tropes*
The importance Kepler gave to the regular polyhedrons cannot be explained apart from his interest in Platonic philosophy and Pythagorean hermeticism. The *Timaeus* nourished centuries of speculation on those geometric solids. They were discovered by Pythagoras, who was also the source, according to *De placitis philosophorum* (attributed in the sixteenth century to Plutarch), of the symbolic meanings attributed to them.[14]

In *Mysterium cosmographicum* as well as *Harmonices mundi*, Kepler recalled the Pythagorean and Platonic association between the five regular solids and the five elements. Here are some of the "resemblances" that formed the basis of the association:[15]
• The cube, rising from a quadrangular base, gives an impression of stability, a quality that belongs equally to the earth.
• The octahedron, suspended from two opposite angles and turned as on a lathe, presents an image of great mobility; it can be compared to air, the most mobile of the elements.
• The tetrahedron suggests fire with its pointed appearance and its sharpness.
• The "globular" form of the icosahedron, which has the great-

est number of surfaces, corresponds to the form of a drop of
water.

• The dodecahedron, containing the same number of surfaces
as the number of signs in the zodiac, and enclosing the greatest
volume, corresponds to celestial matter.

Even while he obligingly mentioned these traditional comparisons, Kepler distanced himself from them. As ingenious as
they are, they nonetheless contain an element of arbitrariness. If
the icosahedron is associated with water, and the dodecahedron
with celestial matter, they both could just as well, and in terms
of "resemblances" that are equally valid, be compared to fire (GW,
vol. 6, p. 81). Other meanings are possible as well. The number
of the surfaces and angles of each polyhedron can be related to
the psychological character associated with each planet (GW,
vol. 1, pp. 34–36). Even a sexual symbolism can be established:
the cube and dodecahedron can be taken as "masculine" solids,
whereas the icosahedron and octahedron have a "feminine" appearance, and the tetrahedron is androgynous. It is even possible to conceive of the cube as "married" to the octahedron, and
the dodecahedron to the icosahedron, while the tetrahedron plays
the role of mediator in the first of these two marriages (GW,
vol. 1, pp. 292–93).

Kepler recalled and commented on these analogies in detail,
with obvious pleasure. But he also disowned them. It is as though
he could express a profound need to meditate on qualitative
resemblances, only if he denied the need. In the second edition
of *Mysterium cosmographicum*, the chapter devoted to the analogy
between the planets and the five regular solids is accompanied
by a note that turns the discussion into a game: I do not say what
I say....[16] On the other hand, the first chapter of the last book
of *Harmonices mundi* treats in detail the genealogical and famil-
ial metaphors, the theme of the sex of the polyhedrons and their

marriages. Thus his attitude toward qualitative similarities is ambivalent: the game is a game, but he plays it diligently. In a similar spirit, Zuccari dedicated a long chapter in his *Idea* to the explanation and etymology of the word *disegno*. He marshaled a variety of considerations intended to prove the divine origin of *disegno*, which is a "sign of God in us" [*segno di dio in noi*]. But in the guise of a transition, he says in the middle of the chapter, "*But to joke a little more about this name...*" [*Ma, per scherzare ancor un poco intorno a questo nome*].[17] Certainly, as Michel Foucault writes, "The age of resemblance is drawing to a close."[18] But the game, even conceded to be arbitrary, is not abandoned without nostalgia and ambiguity.

Kepler attempted to "save" the analogy between the five solids and five elements by relating it to his own system. If Aristotle ridiculed this association (*De caelo* 3.8), it is perhaps because he failed to take into account the arcana with which the Pythagoreans shrouded their mysteries. It is possible that the analogy should not be retained in its obvious, currently accepted sense, where each solid is equivalent to an element; perhaps, the elements can mediate a relation between solids and the planets:

What if the Pythagoreans taught what I teach, but concealed the meaning behind a veil of words?... For if the arrangement of spheres was the same for the Pythagoreans as for Copernicus, if the five bodies and the necessity of their limitation to five were known, if they had all consistently taught that the five bodies are the archetypes of the parts of the universe, what more would we need to believe that Aristotle had only read and refuted the literal sense of their enigmatic thinking? (GW, vol. 6, pp. 17–18)

The Pythagoreans had the habit of concealing their knowledge behind rhetorical veils. To accept the association of polyhedrons and elements in its literal sense is to ignore their usual discursive practice. Kepler continues:

> When Aristotle read "the earth," with which they associated the cube, perhaps they understood "Saturn," whose sphere is separated from Jupiter by the interposition of a cube. And, in effect, we now attribute a state of repose to the earth, and Saturn is endowed with a very slow motion, close to repose, such that the Hebrews named it after repose. In the same manner, Aristotle said that the octahedron is associated with air, whereas perhaps they understood Mercury, whose sphere is inscribed in an octahedron; and Mercury is no less rapid (yes, of all the planets, it is the most rapid) than mobile air is held to be. With the word "fire," perhaps they were alluding to Mars, which in another time took from fire the name of Pyroïs; and the icosahedron was attributed to it, perhaps because its sphere is inscribed in this figure. And under the guise of "water," with which the tetrahedron is associated, the star of Venus (whose sphere is inscribed in an icosahedron) could be hidden, because liquids are subject to Venus and the goddess is said to have been born from sea foam (whence the name Aphrodite). Finally the term "world" could designate the earth, and a dodecahedron could be assigned to the world because the earth's path is inscribed in this figure.... And so in the Pythagorean mysteries the five figures were not correlated to the elements, like Aristotle believed, but to the planets themselves.

Aristotle had mistaken the veil for what lies beneath. The relations *in verbis* that he refuted concealed another relation *in rebus*. For the Pythagoreans themselves, the correspondence between solids and elements was an *involucrum* (a classical term in allegorical exegesis): "they wove meaning with the wrapping of words" [*sententiam involucris verborum texerunt*]. They concealed the constructive function that they assigned to the polyhedrons in the creation of the universe, a function rediscovered by Kepler himself. This is a plausible hypothesis, since we correctly attribute to some Pythagoreans a belief in a heliocentric universe, the only model to contain six planets and thereby to realize the necessary conditions for the embedding of the five solids. This in turn leads to the reconstitution of a complex semantic operation, superimposing three relations. In the case of the cube and Saturn:

| Relation | Substitution | Motivation | Figure |
|---|---|---|---|
| *in verbis* | cube/earth | similarity | metaphor |
| *in verbis* | earth/Saturn | similarity | metaphor |
| *in rebus* | Saturn/cube | container-contained | metonymy |

It is possible that Aristotle did not understand the true meaning of the Pythagorean association. The analogy can be "saved" if it is understood as *veiled*.[19] From such a perspective, the *Mysterium cosmographicum* continued the work of reconstitution begun by Copernicus in *De revolutionibus*. Following the Copernican restoration of heliocentrism already preached by Aristarchus of Samos and others, Kepler restored the constructive role of the five regular solids. In both cases, the teaching of the Pythagoreans is revived. But whereas the correspondence between the Copernican system and the ancient Pythagorean system was obvious, Kepler devoted himself to a *recoding* of a customary reading of an ancient theory, bringing his own innovation, his own difference, to the imitation.

## Mathematics and Marquetry

### Cosmological Mathematics

The theme of embedding the five regular polyhedrons in a sphere, or the inverse, was well rooted in geometric tradition. Euclid devoted the last book of his *Elements* to it. Two supplementary treatises on the regular solids were added to the *Elements* by Hypsicles. Foix de Candale added still another on the same subject.[20] Thus we end up with a total of sixteen books, regularly presented as a single whole, and bestowing particular prestige on the treatment of polyhedrons. Because of its placement at the end of the *Elements*, it could be seen as the crowning glory of the Euclidean enterprise. For Henry Billingsley, the regular solids "are as it were the ende and perfection of all Geometry, for whose sake is written whatsoever is written in Geometry."[21] From a cosmogonic perspective, Proclus had already noted that by culminating in the regular solids, the *Elements* leads to precisely those figures which according to Plato constitute the original forms of creation.[22] Thus it was possible to find in Euclid's plan not only the product of rigorous deduction, but also a teleological orientation leading from geometry to cosmology.

From the cosmological point of view, Kepler attached much importance to the paradigmatic contrasts between the five solids. These contrasts account for the order in which they appear.

The regular polyhedrons are divided into primary polyhedrons and secondary polyhedrons: the cube, the tetrahedron, and the dodecahedron are primary; the icosahedron and octahedron are secondary. The secondary polyhedrons are generated from the former, because their angles include more than three sides, and so on. Among the primary solids, the cube is first in rank: it alone is entirely dominated by the relation of equality; in addition, the tetrahedron is generated from the cube by subtraction, and the dodecahedron by addition. Thus the cube occupies the outermost sphere of the solar system, followed by the two other primary solids. The secondary solids are generated from the dodecahedron and the tetrahedron. The octahedron ranks first among these; hence it occupies the innermost position, just as the cube, the first of the primary solids, occupies the outermost position. The earth occupies a remarkable position in the whole, since it is surrounded by the three primary solids and encloses the secondary solids. It would be tedious to review in detail the diverse and precise reasons for this order of succession.[23] For Kepler, nonetheless, the matter is serious: after *Mysterium cosmographicum*, he returned to it at length in his *Epitome* (4.3) and more briefly in *Harmonices mundi* (5.3). The most important thing to note is that Kepler made an attempt to differentiate the polyhedrons and place them in an ordered classification depending on the investigation of the symbolic meaning of the sphere and the contrast between the curve and the straight line:

– Why do you call these five [figures] the most beautiful and perfect?
– Because they imitate the spherical image of God as much as a figure formed with straight lines can do, because they arrange all their angles in the same spherical volume and they can be inscribed within a fixed volume. And since the surface of a sphere is everywhere equal, the surfaces of each of these figures are equal to each other, and all can be inscribed in a single circle, all their angles being equal. (GW, vol. 7, p. 272)

In contrast, moreover, to the mathematical treatment of polyhedrons, Kepler's model achieves the simultaneous nesting of all the solids. What in the *Elements* is *developed* in a series of theorems, Kepler *enveloped* in a single cosmological construction. In terms borrowed from Nicholas of Cusa, one could say that the universe is the *complicatio* to which geometry offers the *explicatio*. Kepler's geometry is always faithful to the Platonic framework, "suited to draw the soul in the direction of truth, suited to perfect philosophical thinking."[24]

*Abstraction and the Poetic Conceit*
The significance that Kepler gave to the regular solids can doubtless be explained primarily by his belief that God is a geometrician, by his interest in Pythagorean and Platonic speculations. But this preoccupation possesses other intertextual connections. The study of the five solids played a central role in the development of pictorial perspective. Piero della Francesca, Pacioli, and their successors granted the highest importance to the projection of the regular polyhedrons on a plane. This projection was in a way the outcome of the technical education of the painter, as can be seen in the treatises of Dürer or Cousin. The treatment of polyhedrons is the culmination of their work, as it was for Euclid.[25] Following pictorial theory, we should recall the practice of "geometrism" mentioned above. Whenever a purely formal stylization came into play in the theoretical argumentation and preparatory sketches of a Renaissance painter like Dürer, it fused with other requirements in the completed works. For Mannerist artists like Schön and Bracelli, however, the reduction to geometric form became an end in itself.[26] The Germans Jamnitzer and Stoer composed entire landscapes "where the vegetative world is replaced by disturbing whorls, and where polyhedrons and globes in unstable equilibrium replace the human figure," while many other artists "present strictly geometrical objects with loving care."[27] The complete title of Jamnitzer's work on polyhedrons clearly links Euclid and Plato: *Perspective of Regular Solids. This is a detailed description of the five regular solids about which Plato writes in*

*the* Timaeus *and Euclid in his* Elements. [*Perspectiva corporum regularum. Das ist eine fleyssige Fürweisung wie die fünf regulierten Cörper darvon Platon im* Timaeo *und Euclides in sein* Elementis *schreibt*].

By assigning polyhedrons a structural role in the cosmos, Kepler renewed a cosmological tradition in the mode of an artistic tradition. On the one hand, polyhedrons had fulfilled a cosmogonic function ever since the Pythagoreans; but the Pythagoreans had associated polyhedrons with the five elements, without providing them (at least explicitly) with a role in organizing space. On the other hand, in art these same polyhedrons played a structural role in the treatment of perspective space and in the stylization of objects. Kepler's accomplishment in *Mysterium cosmographicum* is to have transformed the cosmogonic function of the five solids by transferring it from the creation of matter to the construction of cosmic space.

By his attachment to the idea of a Creation modeled on polyhedrons, Kepler also brings to mind a painter like Uccello, for whom artistic creation was inseparable from a veritable obsession with geometric solids. Vasari reported an anecdote according to which Donatello reproached Uccello with spending too much time on geometric investigations, in particular the study of polyhedrons "spheres showing seventy-two facets like diamonds" [*palle a settantadue facce a punte di diaminti*] and "other oddities." "These things are no use," he said, "except for marquetry...."[28]

A preoccupation suitable for those who do marquetry.... Marquetry, or inlay work, was a major art, closely linked to the development of perspective. If perspective presupposed the geometric nature of representational space, marquetry consisted of constructing representations by assembling geometrically regular shapes. Marquetry implies a "poetics of abstraction." Through marquetry, "speculative pleasure begets a style," and "mathematical form creates its object."[29] These formulations also apply to Kepler's thesis in *Mysterium cosmographicum*: aesthetic principles take on an abstract character, speculation blossoms into stylistic pleasure, and geometric archetypes generate the form of the cosmos.

Galileo, for whom Kepler's work seems to have provoked so few echoes, launched a critique of Tasso in which the extension of the poetics of marquetry into other domains appears as one of the principal aspects of Mannerism. Galileo reproached Tasso for filling his verses with conceits which are not "necessary" to the plot and which shimmer with meanings that seem grafted onto the subject rather than unified with it. He charged that reliance on such procedures shows poverty of invention: "Since he often has nothing to say, he is obliged to patch together unrelated and disconnected *concetti*." The veneer of conceits he creates is comparable to an inlay: "this filling of stanzas with *concetti* lacking any necessary continuity with things said or to be said, we call marquetry."[30]

Conceits serve to fill a void (*empiare le stanze*). Literally, it is the *voids* in Creation that Kepler sought to "fill" by embedding polyhedrons in spheres. In *Mysterium cosmographicum*, he set out to articulate the universe as a work without flaw, without any *gaps*. Could the manner in which empty space is filled depend on the organization of a conceit? For some critics, it is a fact that Kepler's thesis about the regular solids seemed a useless invention, a sort of gratuitous veneer. Praetorius was one such critic:

But this speculation on the regular solids, what does it add, I ask, to Astronomy? It can be useful, he says, to define the order or the size of the celestial spheres. But it is clear that the distances can be derived from another foundation, namely observation, and *a posteriori*. As for the distances he defines, what difference does it make if they are in agreement with the regular solids? (GW, vol. 13, p. 206)

Moreover, Kepler based the order of the polyhedrons on a *genealogical* metaphor that links their "nobility" to their more or less direct descent from the sphere and permits them to be sexually differentiated and to enter into marriages. All this concern with genealogy and titles of nobility – which would also be directed to the polygons[31] – led Galileo to treat the metaphor of family and social rank in an ironic *concetto*: "it is not I who want the sky to have the noblest shape on account of its being a most noble body; it is Aristotle himself.... As for me, never having read the pedigrees and patents of nobility of shapes, I do not know which of them are more and which less noble, or which more and which less perfect. I believe that all of them are, in a way, ancient and noble; or, to put it better, that there are none which are noble and perfect or any that are ignoble and imperfect, except in so far as for building walls the square shape is more perfect than the circular, while for rolling or for moving wagons I deem the circular more perfect than the triangular."[32]

### Toward the Oxymoron

In spite of new and more precise observations, and in spite of the introduction of a "celestial physics" predicated on the ellipse rather than the circle, Kepler never completely abandoned the belief that the geometric organization of the solar system is based on the five regular polyhedrons. At the very most, he was willing to reduce the importance of the polyhedron when he was ready to provide a firm foundation for a complementary harmonic order.

Meanwhile, he even envisaged subsidiary applications for his theory. When Galileo discovered Jupiter's four moons, Kepler advanced the hypothesis of a localized system of nested polyhedrons. The relation between the orbits of those moons could in effect, he thought, be expressed by three semi-regular solids (GW, vol. 4, p. 309).

In addition, both before and in *Harmonices mundis*, he attempted to adapt his general theory to the facts by turning to other types of polyhedrons. For if Euclid had proved that there are only five such bodies, his proof was only valid for convex polyhedrons. Kepler discovered two other polyhedrons that possess all the qualities of the five Pythagorean solids except convexity. These solids, the small and large star-shaped dodecahedrons, he called the hedgehog (*echinus*) and the oyster (*ostrea*).[33] In 1809, Louis Poinset discovered two other polyhedrons of the same type, and two years later Augustin-Léon Cauchy demonstrated that this class could include a total of only four solids. Kepler is also credited with discovering the rhombotriacontahedron and the octagonal star. But these had already been represented in art. Pacioli had drawn the octagonal star, which he called the "raised octahedron." He had, moreover, imagined figures very close to the two regular solids discovered by Kepler. These solids also appeared as ornamental forms in printing,[34] as well as in treatises by Jamnitzer and Barbaro, another confirmation that Kepler's preoccupation with polyhedrons is closely linked to artistic investigations.[35]

But Kepler also tried to put his discoveries to use in his cosmological speculations. Above all, the *echinus*, the small star-shaped dodecahedron, held his attention. The *echinus* corresponds to the figure obtained by constructing pyramids out of five isosceles triangles on each surface of a regular dodecahedron. Kepler believed that the *echinus* was as closely related to the dodecahedron as to the icosahedron. It was related to the first by its mode of generation, and to the second because of its star-shaped appearance, derived from pyramids erected on a pentagonal base. In addition, the *echinus* was more beautiful than these two other regular polyhedrons: "But the relationship of these figures to the dodecahedron on the one hand and the icosahedron on the other is such that these two, especially the dodecahedron, seem somehow truncated and mutilated when compared with these pointed solids" (GW, vol. 6, p. 82).

The relationship between the *echinus* and both the dodecahedron and the icosahedron inspired Kepler to envisage the introduction of the new polyhedron into his world system. The dodecahedron, which he associated with Mars and the earth, and the icosahedron, which he supposed to be embedded between the earth and Venus, were in effect the most difficult polyhedrons to maintain in the face of the numerical data. Now if it is true that the earth's path – separating the dodecahedron from the icosahedron – is the "measure of all the others" (GW, vol. 1, p. 13), and therefore is not necessarily embedded in any particular polyhedron, it would seem legitimate to look for a polyhedron that directly links the paths of Mars and Venus and is in closer agreement with the numerical data. The *echinus* was just such a candidate (GW, vol. 6, p. 299): "if the augmented dodecahedron that I have called the *echinus*, which is constructed of twelve five-angled stars and is very close to the five regular bodies, if this body places its twelve points on the interior circle of Mars, the pentagonal sides...reach the mean circle of Venus." A localized application that employs the semi-regular solids, turning to the concave polyhedrons with the expectation that they will eventually replace the convex polyhedrons: Kepler's aim, as we have already said, is to eliminate all gaps in the plan of Creation, to inform all relations with geometric reason.

This perseverance in the search for a geometric ideality behind the organization of the universe might be astonishing insofar as Kepler, beginning with the *Astronomia nova*, represents a turning point in the history of astronomy. Without a doubt, the introduction of this "celestial physics" determines Kepler's importance in a "positive" history of astronomy, a history teleologically oriented toward scientific progress. But to take an interest only in that aspect is to refuse to consider Kepler's specific *text*, or to retain only fragments from it and organize them from a point of view that could not have been his. As H. Tuzet correctly writes: "He seeks to construct a classical edifice. It so happens that he runs up against unequal motion, elongated orbits. His laws are a sacrifice to the facts, to observation, to accuracy."[36] We cannot determine Kepler's place in the culture of his own time without taking into account this marriage of opposites. This same point will arise again during our examination of Kepler's thinking about the ellipse.

CHAPTER NINE

# Thinking the Ellipse

Because of the law affirming that the planets travel in elliptical orbits, Kepler's name has often been associated with the Baroque. Eugenio d'Ors writes:

> When in the interest of reducing the entire universe to a system of lucid laws, Kepler denounced the conception of the ancients, according to which the stars move in a circular field, as too restricted, and proposed another scheme whereby the standard is a more complex curve – the ellipse with its two foci – was he not stylizing astronomy, and in a non-classical but resolutely Baroque manner?[1]

In a similar vein, Severo Sarduy declares that for Baroque art and literature, as well as for Kepler, whose first law is emblematic, "the master figure is no longer the circle, with a unique center – radiant, luminous, and paternal – but the ellipse, which in contrast to this obvious focus adds a second focus, equally active, equally real, but sealed off, dead, nocturnal, the reverse of the solar *yang*, fructifying and absent."[2]

For others, however, the ellipse is primarily a Mannerist phenomenon. This opinion is shared by Otto Benesch, Gustav Hocke, and others.[3] Erwin Panofsky declares that the ellipse "was as emphatically rejected in High Renaissance art as it was cherished in Mannerism."[4] And so the role Kepler assigned the ellipse has given birth to the hypothesis, proposed by Panofsky and upheld by Alexandre Koyré, of an astronomy that appeared Mannerist to Kepler's contemporaries, most notably to Galileo.[5]

For some, it is true, the Baroque is only a particular instantiation of an ever-recurring Mannerism, so the choice between the two is almost of no consequence.[6] But the risk is great, if we identify the Baroque as a Mannerism, that we will end up aligning too many phenomena under a single notion, thereby sacrificing its coherence and efficacy.[7] The ellipse furnishes a good illustration of the need for a diversified terminology. It is appropriate to variegate our symbolic and aesthetic thinking about the ellipse; otherwise, we end up with an idealized figure out of touch with the reality of the ways in which it has been put to use.

## The Mannerist Ellipse

### Narrative Motive

Kepler formulated his first law in *Astronomia nova* (1609), published in Prague, the last great center of European Mannerism, during the reign of Rudolf II.[8]

*Astronomia nova* does not reveal a teleological orientation toward replacing the circle with the ellipse. It narrates the abandoning of the circle as much as it does the establishing of the ellipse. The ellipse appears not as an end, but as a result.

It is significant that the book effectively takes the form of a *narrative*, of the story of Kepler's investigations, and does not propose a systematic unfolding of a truth that is postulated or already known. The author explains why he adopted such an unusual form for an astronomical report.[9] If he chose to write the "historical account of his discoveries," it was not only to introduce the reader to particular knowledge or hypotheses, but also to make him aware of the manner in which the author came to his conclusions, "whether by proofs, or doubt, or even, at the beginning, fortuitously" (GW, vol. 3, p. 36). The autobiographical narrative, presented in terms of both a quest and a war,[10] was able to incorporate the distractions, the hopes, the disappointments, the frustrations, the false starts, the role of chance, even episodic digressions. Such a procedure vindicates the conclusions of *Astronomia nova*: the reader will see that they do not arise from a desire to put an end to the dogma of circularity or to innovate at any cost, but on the contrary, from an odyssey undertaken under the very sign of the prevailing belief in the circle. If the hero-narrator "reports many new ideas," he shows that he "has done so *under constraint*" (GW, vol. 3, p. 36). The ellipse is a constraint, not a choice.

The very first lines of the first chapter recall with undiminished respect the nobility of the circle and its suitability in astronomy:

> The changelessness of the motions of the planets shows that they are circular. This fact having been established by observation, reason presumes that their paths are perfect circles. For we judge that the circle is the most perfect of figures, heaven [the most perfect] of bodies. (GW, vol. 3, p. 61)

Kepler's first hypothesis for defining the exact path of Mars relied on traditional methods: preserving circular motion and even reintroducing the equant, which Copernicus had abandoned.[11] Only at the end of Book Two does Kepler's doubt about circular motion explicitly surface:

It is necessary therefore that there is something false in what we have supposed. It was supposed that the orbit traveled by the planet was a perfect circle and that on the line formed by the apsides there was a unique point, at a precise and constant distance from the orbit around which Mars traveled equal angles in equal time. One of these supposition is incorrect, perhaps both. For the observations are not false. (GW, vol. 3, p. 176)

And it is not until the end of the third of five books that Kepler definitively affirms the necessity of abandoning the circle and leaving familiar paths behind: "My first error was to have postulated that the path of a planet is a perfect circle; it was all the more pernicious since it was upheld by the authority of all the philosophers and seemed appropriate from the metaphysical perspective" (GW, vol. 3, p. 263).

As soon as he accepted that he had no choice but to abandon circularity, Kepler found himself in what has been called a "no-man's-land" or a state of "absolute intellectual isolation."[12] The perfect figure having been eliminated, all other figures were equally valid candidates. Only after much floundering did he finally discover that the orbit is a "regular ellipse."[13] To those, like his friend Fabricius, who continued to think in terms of the circle, he now responded that their constructions rested on nothing real and were only intellectual constructs:

But if you say that it is not doubtful that the motions follow the path of a perfect circle, I reply that for composite motions, that is *for real motions*, this is false. In fact, according to Copernicus they follow, as I have already said, a trajectory that bulges on the sides, and even on the spirals according to Ptolemy and Brahe. If you speak of the constituents of motions, you speak of something that *is only thought, and that in reality is not there.* For nothing completes orbits in the heavens but the bodies of the planets themselves; there is no orb, no epicycle.... (GW, vol. 16, pp. 14–15; my italics)

"Real motion" contrasts with motion "that is only thought." The law of the ellipse represents a submission to facts far removed from a dreamed of perfection. We find the same submission and the same distance among the Mannerists, who verify in their own domain the impossibility of basing a realist representation on codes presumed to be perfect. Mannerism, as we know, is characterized by elongations, foreshortening, and the other deformations to which it subjects the geometric perfection of classical proportions. Among others, Vincenzo Danti devoted a large chapter to the impossibility of following "any measure of quantity perfectly" [àlcuna misura di quantità] in the arts:

It is true that some of the ancients and moderns have written very diligently about the representation of the human body. But it has been established that those works manifestly cannot serve. They wanted, by means of quantitative measure, to establish a rule; but this measure is not perfectly realized in the human body....[14]

In the same spirit, Borghini wrote:

It is necessary to know standard proportions; but we must consider that we are not always in a position to observe them. In effect, we often represent figures in the act of bending, rising, or turning. In these attitudes, the arms stretch or bend in such a way that, in order to make the figures graceful, it is necessary to increase the proportions in one part and decrease them in another. [15]

Increase or decrease the proportions: this is what Kepler was forced to do to circular perfection in order to account for the reality of the planetary orbits. "The orbit of a planet is not a circle," he wrote, "but it curves in little by little and then moves back toward the fullness of the circle..." (GW, vol. 3, p. 286). Like the Mannerists, he accepted forms that depart from established norms, whether philosophical, scientific, or aesthetic.

To be sure, the precise concrete motivations are not the same for artists as for Kepler. But we come across the same type of situation in both cases: acceptance of the contradiction between the ideal and the real, the refusal to distort or ignore what is real.

*Form and Force*

Comparisons between Kepler's ellipse and the artist's ellipse often concentrate on the fact that an ellipse has two foci: we saw this in the quotations from d'Ors and Sarduy. It is nonetheless important to consider models for physically generating and/or symbolically interpreting the ellipse that favor other elements or relationships.

It appears clear that the contrast between the two foci played *no* role in the discovery of Kepler's first law. In his study of its genesis, Curtius Wilson writes:

> Given that the correct orbit is a sun-focused ellipse, one might imagine that Kepler had been led to the idea of this orbit by the appropriateness of giving to the sun a position of such geometrical prominence as the focus.... There is no evidence that Kepler, on surveying the results of the distance determinations and their range of probable error, hit on the

idea of the focus as a clue to the selection of the correct, sun-focused ellipse.[16]

The history of the discovery of the elliptical orbits cannot be explained in terms of the two foci. It is the history of a circle that is deformed, of a center that is decentered. Kepler stands out by his refusal to dissociate thinking about form and thinking about force. Before him, astronomy had been kinetic, establishing an often complex arrangement of circles, but always adhering to the metaphysical insistence on perfect form. Kepler, on the other hand, wanted to provide a *dynamic* explanation for planetary motion: form is a function of force. The sun propels the planets by a *species immateriata* that it generates and the attraction-repulsion of a magnetic force. The point of reference for the planet's path is the sun. A dynamic relationship between the sun and the planet creates a resultant that is both centrifugal and centripetal, effectively decentering the sun. The generation of the ellipse is conceived in terms of the *species immateriata*, of attraction and repulsion, not in terms of the geometry of the ellipse with its two foci. Force creates form. As Kepler wrote to Fabricius: "The difference lies in the fact that you employ circles and I employ corporeal forces" (GW, vol. 16, p. 15). At first consideration, Kepler seems to have brought about a complete transformation from the formistic perspective, dominated by faith in final ends and ideal perfection, to a contextualist perspective, where relations are conceived as interactions between phenomena at the same level of reality.

This conception of the origin of form has its counterpart in Mannerist art. Here too, we see a distancing from Apollonian sufficiency, from the auto-teleological realization of forms in their geometric ideality.[17] Danti remarked that the body lacks "stable proportions" because it is "in motion from beginning to end."[18]

The form that motion takes is not defined as dependent on a stable and stabilizing model, on an ideal conceptual measure. Likewise, Kepler did not conceive of the rotation of the planetary bodies around a purely geometric center:

> I do not deny that one can conceive a center and a circle around the center. I only say that if this center exists only in thought, and is never marked by an external sign, it is not possible for a real body to execute a perfectly circular motion around it. (GW, vol. 3, p. 258)

This impossibility also applies to the human body: motion, the effect of muscular action, never follows the path of a perfect circle. "There is no member that can undergo a uniform and easy circular motion.... But certainly, if there had existed a way to employ the motor faculties so as to produce circular body motion, it would not have been neglected in the human body" (GW, vol. 3, p. 69).

Mannerist art also imagined motion as a function of dynamic relations; form, then, is an effect produced by these relations. The pyramid or conic form, compared by Lomazzo to the "flame of fire" and expressing the "fury of the figure," denotes an energetic relation, an attraction. Both the "flame of fire" and the "pointed cone" seek to split the air and climb to their proper sphere.[19] The serpentine figure, often inscribed on a pyramid and developed from the classical *contrapposto*,[20] implies the decomposition of the body into antithetical pairs (head-shoulders, arm-hand, palm-fingers...), which are attracted and repelled in opposite directions. Aretino interpreted this magnetic structure in dynamic terms – *forza facile* [docile force], *sforzata facilitade* [forced ease] – and attributed to it a special "magnetism."[21] Benesch interpreted some of El Greco's paintings in magnetic terms: transparent lines of force traverse these paintings, mystical visualizations of a *species immateriata* traveling through space.[22]

It is true of course that El Greco's lines of force possess a primarily spiritual meaning. But if Kepler thought of form as the product of a material dynamic, it is no less true that the dualities so characteristic of Mannerism[23] provide form with symbolic meaning and aesthetic justification. In other words, contextualism is interpreted in turn from a formist perspective.

### Metaphor and Oxymoron

As early as *Mysterium cosmographicum*, Kepler was led by symbolic considerations to attribute a dynamic quality to the sun. Copernicus had already superimposed the symbolic motif of the sun (as ruler, king of heaven, visible representative of an invisible God) on the mathematical thesis of its central position in the universe. But as we noted, the sun remains passive: it plays no physical role; its sole function is to justify a visual representation. For Kepler, on the other hand, the distinction of the three persons of the Trinity and the designation of the sun as symbol of the *Father* led him to invoke another motif: the center as dynamic focus, as principle of generation[24]:

> Thus the sun, located in the middle of the moving [planets], itself at rest, yet a source of motion, carries the image of God the Father as Creator. For what the creation is to God, motion is to the sun. (GW, vol. 13, p. 35)

The sun's *physical* role therefore agrees with its symbolic meaning. Kepler adds that the sun "moves in the fixed stars like the Father in the Son. For if the fixed stars did not furnish by their fixity the place [with respect to which motion is perceived], nothing could move." The Father creates *in* the Son. Moreover, he creates motion *through* (across) the space between the center and the surface, just as the Father creates *through* the Spirit. "In effect, the Sun diffuses the force of motion through the intermediate space containing the moving planets, just as the Father creates through the Spirit or the Spirit through its own force." The Latin text plays on the ambiguous repetition of the preposition *per*: "Dispergitur Sol virtutem per medium...sicut pater per spiritum." From the beginning, the sun constituted a particularly determined element: not only the mathematical center (for the calculation of motion) and metaphysical center (of the Father), but also the physical center (the cause of motion).[25] Jürgen Mittelstrass notes that "in the framework of this undertaking, the metaphysics of nature engendered a physical methodology."[26] Hans Blumenberg highlights the importance of the "*medium* of the metaphor: to obtain Kepler's laws, the schema representing the sphere as radiation from the center was more essential than an exact conception of gravity."[27]

*Astronomia nova* acknowledged that the ellipse "seems to possess less geometric beauty" than the circle (GW, vol. 3, p. 181). But it is not sufficient to conceptualize one in terms of the other. With and around Kepler, new reflections on the continuum of conic sections begin. In an analysis of conic sections, the circle and ellipse are not simply *different*; they are *opposing* members of a whole that also includes other forms. In terms of this system, elliptical motion does not destroy, in Kepler's eyes, the perfection of the universe as a meaningful message. The ancients themselves had been obliged to admit the eccentricity of the planets' motion, something that "was a much greater deformity than the ellipse" (GW, vol. 7, p. 331). In addition, the hyperbola, the parabola, and the ellipse are symbolically mixed, participating in both circular and rectilinear motion: thus they mediate the terms of the fundamental opposition between the curve (representing the Creator) and the straight line (representing the creature), which beginning with the *Mysterium cosmographicum* (1597) directed Kepler's conception of Creation. Of the three, the ellipse is closest to circular perfection:

> Thus there exist two opposite boundaries, the Circle and the straight line: here the pure curve, there pure straightness. The Hyperbola, the Parabola, and the Ellipse are intermediate and participate in both the straight line and the curve; the Parabola participates equally; the Hyperbola participates more in straightness, the Ellipse more in the curve. From this fact, the Hyperbola resembles more and more a straight line, i.e., its Asymptote, the further it extends. The Ellipse tends toward circularity to the degree that it extends beyond its middle and finishes by turning in on itself.[28]

Mixture of curve and straight line, the ellipse signifies the submission of the creature to material necessity and its inability to attain total perfection:

> For if it was only a question of the beauty of the circle, the spirit would decide with good reason for it, and the circle would be suitable for all bodies, principally for celestial bodies, since bodies participate in quantity, and the circle is the most beautiful form of quantity. But since it was necessary to rely not only on the spirit but also on natural and animal faculties to create motion, these faculties followed their own inclination, and they were not accomplished according to the dictates of spirit, which they did not perceive, but through material necessity. It is therefore not astonishing that these faculties, mixed together, did not fully reach perfection. (GW, vol. 7, p. 330)

The planets do not *fully* (*penitus*) attain perfection of motion. Let us understand, however, that the ellipse, through its particular affinity with the circle signifies the effort of the creature to resemble its Author *as much as possible.*[29]

Such a valuation, however, cannot be explained solely in terms of the geometry of conic sections. According to this geometry, the circle is a particular case of the ellipse: "of all ellipses... the most open is the circle."[30] In Pascal's terms, all conic sections are only parallel *images* of one other, without any hierarchy, transformations approaching a limit along a mathematical continuum.[31] If greater "perfection" is attributed to the circle on the symbolic level, it is because of the interaction of the geometry of the sphere with the geometry of the cone. As a figure for the Trinity and the form of the universe in its totality, the sphere is, from the point of view of "metaphysical" geometry, first among all forms, whereas the circle is first in honor among the two-dimensional figures that Kepler calls "knowable," meaning they can be constructed with a ruler and compass. In this metaphysical geometry, the ellipse is incontestably a figure of an inferior order, a mixture of curve and straight line that cannot be constructed with a ruler and compass. The ellipse can be produced by cutting a cross-section of the cone or mechanically, with two lengths of string.[32] In a geometry of the line and compass, the ellipse is an irrational figure. Now irrational proportions, supposing an infinite mediation, define precisely, as Kepler said with respect to the squaring of the circle (GW, vol. 1, p. 23), the relation of the creature to its Creator.

Starting with the valuation of the ellipse as marriage of curve and straight line, and thanks to the interaction of the geometry of the cone and sphere, it becomes clear what rhetorical figure the ellipse is equivalent to. One would be wrong to be guided too strongly by the homonym. Since the linguistic ellipsis [*ellipse*] indicates a suppression, it is all too easy to compare the geometrical ellipse [*ellipse*] with those figures of language, that imply a suppression: enthymeme or metaphor. Neither of these comparisons seems to fit Kepler's ellipse. Insofar as it represents the union of contraries, the curve and the straight line, and it is interpreted by consulting two different geometries, it is more like the *oxymoron* – an oxymoron based, of course, on the metaphor of the curve and straight line.

The transition from Ptolemy *and* Copernicus to Kepler implies a shift from thinking in terms of antitheses to thinking in terms of the oxymoron. The Ptolemaic cosmos reserves the circle for the supralunary realm and rectilinear motion for the sublunary. For Copernicus, every totality has a tendency to assume the form of the sphere, whereas motion along a straight line is either a disturbance of order or a return to order.[33] But as soon as celestial motion began to be defined in terms of the ellipse, complementarity became union, and antithesis shifted to oxymoron: the same orderly motion is composed of both curve and straight line. Like the substitution of the ellipse for the circle, the substitution of the oxymoron for antithesis also implies a shift from a kinetic to a dynamic conceptualization: "Only the 'conjunction,' that is the 'coincidence' of antonyms in the nucleus of an oxymoron, places antithetical terms both structurally and functionally in a relation of contradictory dynamism, of more or less direct contradiction...."[34]

In the Renaissance theory of art, Alberti preached the *contrapposto*, according to which "some parts [of bodies] should be shown toward the spectators, and others should be turned away; some should be raised upwards and others directed downwards." But he rejected as "too violent" motions that "make visible simultaneously in one and the same figure both chest and buttocks."[35] This is praise of the non-contradictory antithesis or opposition, rejection of the oxymoron, of the co-manifestation of contraries. The serpentine figure is precisely that coincidence, about which Aretino, in the commentary cited above, said, "it shows the back and front at the same time," and the effect of which he effectively transposed in formulas that are in their turn oxymoronic: *forza facile, sforzata facilitade.*

The concept of the ellipse as a union of contraries is itself the effect of a deeper oxymoron that informs all of Kepler's work. For him, the universe remains a symbolic and coherent message that expresses God's nature and his relation to Creation. Consequently, the universe obeys the metaphoric logic of the anagogical symbol: logic of resemblance between the sign and the transcendence it represents (a transcendence that conditions *a priori* the form of the sign). But to read the universe, it is important that we not let ourselves be guided by arbitrary deductions. We must unveil the real form of the sign, assure ourselves of it *a posteriori* by observation. Tycho Brahe had just added new precision and rigor to observations: a discrepancy of eight minutes between Kepler's original predictions and Brahe's observations was sufficient to cause Kepler to abandon his first (circular) hypothesis concerning the motion of Mars, whereas a difference of that magnitude would have been completely acceptable to Copernicus. All Kepler's efforts were directed toward articulating in juxtaposition and without interruption the contraries of *a priori* and *a posteriori*, deductive logic anchored in anagogical symbols and inductive logic based on empirical facts. The tension is all the greater since the two requirements were expressed with a rigor never equaled. As a "logothete," Kepler developed an extremely constraining symbolic logic of the cosmos; faithful to Brahe's teaching, he refused the slightest deviation from the exactitude of numerical data.

Ellipse and oxymoron: do we need to recall that the oxymoron was also recognized as the dominant figure in Mannerist literature? For Hocke, it is the very indication of the Mannerist personality: "If the oxymoron...is one of the most important stylistic devices of Mannerist poetry, this type of problematic entity appears, in turn, as the personification of the oxymoron, of the joining of contraries."[36]

### Ellipse and Harmony

An aesthetic justification of the ellipse appeared in *Harmonices mundis* (1619), the work with which Kepler hoped to crown his investigations, since in it he proposed a principle that would make it possible to account for the overall construction of the universe without having to renounce the preceding stages of his thinking.

With the ellipse, the relation between form and force now becomes symmetrical. If the elliptical form is still conceived as the effect of a dynamic relation, this relation, in turn, is governed by a formal consideration. A teleological reference is added to the contextual reference.[37] Significantly, the last book of *Harmonices mundis* is entitled "On the perfect harmony of celestial motions, and the genesis *from this harmony* of eccentricities, semi-diameters, and periods of revolution." *From* harmony: this becomes the final cause that explains the ellipse, the dimensions and periods of the orbits. The title of Chapter 9 of this book makes explicit this same orientation toward final causes: "That eccentricities in the individual planets have their origin in the procurement of harmony between their motions" (GW, vol. 6, p. 330).

The circumference of the ellipse is compared to a *cord* looped back on itself, while planetary harmony is realized in relation to the solar focus, the place where the principal body of the universe is located. In terms of Kepler's second law (in equal times, a planet sweeps through arcs of equal area), a planet traverses different fractional parts of the elliptical cord in equal times. Thus, due to variations in angular speed, harmonic intervals are created, and each planet executes a specific musical phrase, while the music of the whole "choir" is polyphonic.

If the musical beauty of the universe is the result of *motion*, it is because motion characterizes the *living*, and God wanted the cosmos to be made in the image of a living being. Now the beauty of the living being implies the laying aside of a purely geometric or geometrically pure perfection:

> Harmonies are on the side of Form, which the hand imposes last.... It so happened that Harmonies gave rise to Eccentricities [i.e., ellipses].... On the statue, so to speak, they gave rise to the nose, eyes, and other parts.... Since neither the bodies of animate beings nor the solid shapes of stones are made according to the pure norm of a geometric figure, but deviate from the exterior round form...such that the body can acquire the organs necessary for life and the stone can receive the image of the animate being; in the same manner, the proportion that the figures prescribed for the planetary Orbs had to give way to the Harmonies, because proportion is inferior and only concerns the body and matter.[38] (GW, vol. 6, pp. 342–43)

The ellipse gives birth to the superior beauty of harmony. Elliptical motion is a geometric imperfection that assures musical perfection: a new oxymoron comes into play in this overlapping of imperfection and perfection. We have seen already that the theoreticians of Mannerist art verified, acknowledged, and even preached deviation from perfect proportions in the representation of bodies and motion. For them also, the important thing was to discover in this deviation a particular beauty. In a passage already cited, Borghini declared that in order "to give grace" to figures we must increase or decrease the proportions.

But the celestial music, as Kepler conceived it, cannot be directly perceived by earthlings. It is apparent only to the sun, and man must reconstitute it by a purely intellectual effort. At best he can know the score; he can never attend the performance. The cosmic concert that crowns the Keplerian system is only imagined by the human mind: it is a musical *idea*.

Doubtless it has always been more or less so, according to the legend that Pythagoras was the only man to have actually *heard* the music of the spheres. But now, for Kepler, that music is an idea at two removes, so to speak, implying a radical divorce between the tangible and the intelligible. In its Copernican form, the heliocentric theory already contradicted perception, requiring the earth and sun to exchange their tangible qualities as body at rest and body in motion. In this purely intelligible universe, Kepler introduced new "deformations," substituting the ellipse for the circle,[39] justifying these deformations by another, musical intelligibility. In sum, the "cosmicality" of the cosmos continues to be guaranteed at the level of the intelligible.

Now the Mannerists interiorized the conception of *gracefulness* in just such a manner. According to Danti, *la grazia* is known "by means of intellectual power." In terms of this principle, the Mannerist theory justifies all the distortions of visual representation to the degree that they make it possible to express an abstract idea more completely. For Kepler, as well, deviations and distortions arise from the realm of appearance and are at the service of a purely intelligible beauty.

### Toward the Baroque

*Stevin*
In 1605, while Kepler was working on *Astronomia nova*, Simon Stevin published a treatise on astronomy in which he proclaimed

the truth of the Copernican system: *De Hemelloop* [*The Heavenly Motions*].[40] Even in the Netherlands, the publication of such an unwavering defense of the Copernican system was a courageous act. The work immediately brought down on its author thunderbolts from the Calvinist Ubbo Emmius, the future rector of the University of Groningen.[41]

Stevin did more than copy or summarize the *De revolutionibus*. He clearly juxtaposed Ptolemy and Copernicus, showed the effectiveness of each, permitting the reader to understand why Copernicus could be preferred, above all stressing the greater coherence of the heliocentric model: "what is considered strange and gives rise to astonishment in the theory of a fixed Earth does not give rise to astonishment in the other theory [of a moving Earth], because it is based on what actually happens in nature."[42]

On the other hand, Stevin, like Kepler, upheld the idea of a kind of magnetism. He is not interested in the ellipse, nor does he ascribe special favor to the sun: magnetic force is not exerted from the center to the periphery but emanates from the sphere of the fixed stars.[43] He does not ascribe formist or symbolic value to the circumference or to the "high" – as in the cosmos of the ancients – since a purely physical force effectively replaces the ancient concept of a metaphysical *primum mobile*. Stevin sets out to provide an explanation that is totally contextualist. His abandoning of formism and its symbolico-anagogical correlates follows cleanly from his commentary on the position of the sun. Stevin stresses that the Copernican sun is not really at the center of the planetary paths, and he is not even sure it is at the center of the fixed stars. Given the immensity of the sphere of fixed stars, the entire solar system is reduced to a point and can henceforth be considered as the center:

As to the world's center [i.e., center of the universe] on the theory of a moving Earth, for that the Sun is justly taken, because it is sufficiently near the center of the circles described through the apogees of the Planets' orbits; but that it is the center of the Heaven of the fixed stars can be surmised, but in my opinion it cannot be fully proved....

As to the fact that *Copernicus* asks in the tenth chapter of his first book who would in this beautiful church place that lamp in any other, better place than in the middle, from where it can illuminate everything: these are moving, natural reasons indeed, but not based on a geometrical proof. So much is true that if one wished to take any point other than the Sun, for example the center of the Earth's orbit, for the center of the world, assuming the Sun to revolve about it, with a semi-diameter equal to the line of eccentricity of the Earth's orbit, one might base on this a description of the Heavenly Motions without any error; but it is more convenient to take the Sun for the center of the world, both in order to learn properly the correspondences of the theories of a fixed and a moving Earth, which are to be described hereinafter, and on account of other matters that may arise, which are easier and more intelligible in this way.[44]

If Stevin believed in the truth of the Copernican system, he refused to conceptualize that truth in the formist and anagogical framework that Kepler still operated within. Certainly, in his praise of the central position of the sun, he acknowledged "naturally pregnant reasons" [*beweeghlicke natuerlicke Redenen*]. But neither the possibility nor the rejection of a symbolic interpretation really concerned him. In the same passage, he compared the choice of the center of the universe to the unconstrained choice of a first meridian in geography: a purely practical convention for the manipulation of signs.

*Galileo*

It is a well-known fact that Galileo, while aware of the law of the ellipse, made no use of it and never even referred to it.[45] Panofsky and Koyré have seen this attitude as evidence of a return to "classical" Renaissance aesthetics, of an anti-Mannerist stance which is also clear in Galileo's commentaries on Tasso:

> While we were considering the circle as a particular case of the ellipse, Galileo could only experience the ellipse as a deformed circle: a form in which "perfect order" had been disrupted by the intrusion of the rectilinear; which, for this reason, could not be the result of what he considered to be uniform motion; and which, we can add, had been rejected by the classical Renaissance and cultivated by the Mannerists with equal fervor.[46]

The pertinence of such an explanation has been placed in doubt, especially by Maurice Clavelin. Far from reflecting a belief in the immanent perfection of a Platonic *eidos*,[47] Galileo's predilection for circular forms may be based on his verification of their efficacy. Since the universe is manifestly directed by a principle of order, and since the circle appears to be the best means of maintaining the stability of that order, the circle's predominant role in nature can be explained primarily by *practical* reasons: "It is completely useless to invoke aesthetic motivations or the so-called intrinsic perfection of circular motion.... The power of circular motion never to corrupt order is amply sufficient to explain its promotion to the rank of a cosmological premise."[48]

We must admit that a motivation related to the contextualist functionality of geometric figures agrees more fully with Galileo's famous declaration in *The Assayer* concerning the "nobility" of forms, where he proclaims that he does not know which

geometric figures are nobler or more perfect, and that he is satisfied to confirm that round figures are better for rolling than
triangular figures, and squares better for building than spheres.[49]
In light of this passage, it seems difficult to accept that an aesthetic deficiency or a symbolic inferiority would be enough to
turn Galileo away from the ellipse. Instead of describing the
ellipse as a "distortion of the circle," *The Assayer* explicitly clas-
sifies it among the "regular figures."[50] A more satisfying explanation of Galileo's silence may be found on the side of efficacy.
Kepler's second law – inseparable from the first – posed huge
mathematical difficulties, since it did not provide for a directly
calculable relationship between the position of a planet and the
period of a revolution, or between the area and the angle trav-
ersed. In his study of the reaction to Kepler's laws, J. L. Russell
notes: "The chief complaint levelled against his planetary theory
was that the area law was 'ungeometrical' and that in order to
use it one had to resort to devices which were unworthy of a
mathematician."[51] As Gerald Holton writes:

> The analysis of planetary motion along elliptical orbits is
> repugnant not only to classical aesthetics but also to the crude
> manual calculator and the impartial computer. The simplest
> approach to periodic motion is to consider it as the result of
> a suitable number of circular (harmonic) motions of differ-
> ent amplitude, frequency and phase. In the last analysis, today
> we use Kepler's diagrams to picture planetary motion, but we
> adopt Galileo's decision to stay with the circle to understand
> the calculation of planetary motion.[52]

We can suppose therefore that practical difficulties, rather than
an indestructible faith in the perfection of the circle, turned
Galileo away from using Kepler's laws. As soon as the task of per-

forming calculations arises, the second law has the inconvenience of looking like the circular brick or triangular wheel alluded to in *The Assayer*.

I am suggesting that Galileo's objection to the ellipse is based on a practical, functional criterion rather than on the Platonic norm of the self-realization of perfect form. Without establishing a causal link (in either direction), we note that this orientation toward the usage value of signs is also found in Baroque aesthetics. The environment in which Galileo evolved was after all Baroque Italy.[53] And if the author of *The Assayer* judged geometric figures in terms of their interpretive utility, without the usual prejudices, the Baroque evaluated forms in terms of their effect on an audience. Mersenne provides an example of the application of Baroque rhetoric to geometry, and especially the ellipse.

*Mersenne*

In *Quaestiones in Genesim*, Mersenne proclaimed that "geometry is useful for expressing more fully God's qualities and works."[54] The affirmation is illustrated by lengthy discussions, some inspired by Nicholas of Cusa, especially concerning the circle. But the perspective is no longer Kepler's. The objective is not to discover an original meaning in geometric forms. Mersenne juxtaposes different comparisons between the circle and God as the outcome of a game with multiple solutions. To move from one comparison to another, he uses such a formula as "if you do not prefer to call the center of all divinity the eternal father...."[55] *Nisi malueris*: no figurative meaning is intrinsic to the circle; all are the products of the human mind which manipulates them. For Mersenne, the circle becomes, in the terms that the Italian Tesauro applied to the human body, "a page always ready to receive new characters and see old ones erased."[56]

The same conclusion applies even more clearly to the third

volume of *Harmonie universelle*, which discusses in detail the figurative meanings of the ellipse and other conic sections. Here
are several examples:

> One can be ignited in divine love by the comparison to coals
> heated by parallel rays at the focus of the parabola, hyperbola,
> or ellipse....
>
> And if, rather than God complaining about his people
> when he says, *Factus sum illis in parabolam* [I became a byword
> to them], we take it in another sense, understanding a refer-
> ence to the conic parabola, it is a parable that will inflame
> our hearts with his love, and a hyperbole that gathers and
> unites our distracted thoughts, *Dispersiones Israel congregabit*
> [He will gather together the dispersed of Israel], and directs
> them to a single point, of which it is said, *Porro unum est
> necessarium* [and yet only one thing is needful].
>
> If we imitate the elliptical concave [figure], we will trans-
> fer the hot coals of love that they have for God from their heart
> to ours, as the ellipse carries all rays from one focus to the
> other, in order to justify the thought of the royal Prophet,
> *Particeps ego sum omnium timentium te*, etc. [I am a partaker
> with all them that fear thee], and we will...even be able
> to appropriate the light of science and the fire that results,
> which they neglect, by imitating the convex hyperbola that
> gathers at the exterior focus that which is derived from the
> interior focus.[57]

There is no more hierarchy, as with Kepler, but alternation.
All the conic sections – like the circle in *Quaestiones in Genesim* –
express with equal validity different aspects of God – or man.
This figurative language is no longer a language of *participation*,
but of *representation*. Mersenne makes no claim that these com-

parisons elucidate a divine language. The citations serve only to illustrate the "usage of mathematics on behalf of the Preachers."[58] At issue is a *usage* of practical, tactical, and didactic finality, representing religious truth to an audience in a form that excites attention by its novelty. Be pleasing and teach. From Kepler to Mersenne, we pass from absolute metaphors to a restrained rhetoric, aiming at varied effects in particular contexts. Mersenne's proposed usage for geometric forms in persuasive discourse is analogous to the usage Galileo assigned them in the study of nature: they are functional, without reference to an immanent metaphysical sense. Such a usage is also found in applying the ellipse to architectural ends. The ellipse performs a precise function in a given context without transcendent meaning: "but we must find the points of the Ellipse that are called the *foci*, since light and sound are reflected from one to the other and produce one of the most admirable effects in all of nature."[59] It is also in such a spirit, envisaging the production of striking contexts and not the formist realization of a figure, that Baroque architects arranged the entrance and the altar, sometimes along the large axis of an elliptical church (San Carlo alle Quatro Fontane), sometimes along the small axis (Sant' Andreo al Quirinale), producing the effect of either contracting or expanding space.[60]

### Tesauro and Pascal

Mersenne's metaphors and comparisons are fluid: the vehicle of comparison is by turns ellipse, parabola, hyperbola. None of these is favored over the others; the metaphor traverses one by one the different sections of the cone. It is, however, consistently enunciated in terms of the cone, which becomes the fundamental figure, generating all others. To be sure, the perspective cone did not come into being with the Baroque. But from the Renaissance to the Baroque, passing through the crucial stage of Mannerism,

reveal a process of representation where the multiplication of perspectives highlights the absence of an absolute perspective.[66]

All we can do is arrange by enumerating and contrasting from our "remote corner of nature." Certainly, in such a practice, the oxymoron remains present. But it changes register altogether. The oxymoron of Kepler's ellipse was of a semantic order: it related to a meaning that could be thought and spoken; man, image of God, participates in the idea expressed in the universe. Pascal's oxymoron is of a referential order: all knowledge is non-knowledge. And, if knowing proceeds by antithesis, it also proceeds by grammatical ellipsis: ellipsis of the failure of the reference, of the relativity of Galileo's notion of functionality.

If the Baroque oxymoron is not the Mannerist oxymoron, the Baroque use of antithesis is also not that of the classical Renaissance. Antithesis and oxymoron are completed in a specific manner. In the antitheses of Baroque poetry, "qualities are organized by differences, differences by contrasts, and the world is polarized according to the strict laws of a kind of material geometry."[67] Gérard Genette discretely compares this reliance on antithesis to the representation of the self-renewing universe: "the goal is to master a universe enlarged beyond all measure, decentered, and quite literally *disoriented*."[68] The parallel between such a formal procedure and Pascal's method is striking, but we know that Pascal condemned the *artificial* order of poetic antitheses that "force words" into "symmetry" with the help of "false windows."[69] If Pascal's analysis is correct, his criticism is perhaps unjust. The "false windows" may be there to mimic the dissolution of any fixed vantage point from which to articulate – the same dissolution that the *Pensées* put to the test. Forced figures lead to Tesauro's ambiguous smile, whose meaning was clarified by Raimondi,[70] and which in its own way is part of what is most profound in the Baroque: connotation, in the trajectory from *inganno* to *disinganno* (deceit to disillusion), of the oxymoron of reference. We should not forget that for Tesauro witty conceits are like the "apples of the Black Sea: beautiful to look at, but tasting of ashes and smoke."[71]

## Descartes

The ellipse and circle appear in Descartes' *Geometry* without the slightest hierarchical distinction. Along with the parabola and hyperbola, both are curves "of the first and simplest type."[72] The distinction between figures that can be drawn with the ruler and compass, and those that cannot has been abolished. All figures that "fall under some exact and precise measurement," that is whose points (and no other) bear the same relation to the x and y axes, are "geometric."[73] The points along any straight line are characterized by the relation $x=y$, those on a parabola by $y=x^2$, and so on. Geometric figures can be translated into algebraic formulas: "Although it is easy to explain them [conic sections] more clearly than Apollonius or anyone else, it is still I think very difficult to say anything about them without Algebra that cannot be said much more easily with Algebra."[74] In the words of Florimond de Beaune, Descartes' *Geometry* shows "the mutual relations and convenience of arithmetic and geometry."[75] Or, as Descartes put it: "The method I use reduces everything that falls within the purview of geometricians to the same kind of problem as looking for the solution to an equation."[76]

Kepler had always distrusted algebraic calculations. He saw a contradiction in the aspiration to "analyze" continuous or geometric quantity with discrete (discontinuous) numbers.[77] For him, algebra was limited to utilitarian calculations. It could not provide the basis for a cosmological order: "I do not treat these matters by numbers or by Algebra, but by the investigation of Spirit; my interest in these matters is not for keeping a ledger but for explaining the causes of things" (GW, vol. 6, p. 19). In cosmology, "the Regular figures are examined for themselves: as Archetypes, they contain their own perfection within themselves..." (GW, vol. 6, p. 18). For Kepler, the formistic conception, according to which some geometric figures and the universe are metaphoric signs of divine Ideas, was still dominant.

If Descartes rejected formistic geometry and the archetypal hierarchy of figures, he nonetheless continued to ascribe a transcendent meaning to mathematics. This meaning is no longer located in the Idea represented by any given figure. There is no longer, for example, an order of *metaphor* that presupposes an analogy between the sphere and the Trinity or between motion toward perfection and the ellipse. Meaning is purely *metonymic*: the perfect and innate knowledge that I have of geometric figures, as a *present effect* in my imperfect being, is the mark of an *absent cause*, of a perfect being: "The human mind possesses something divine, where the first seeds of useful thoughts are sown.... The proof is in the easiest of the sciences, arithmetic and geometry...."[78] Reducing geometry to algebra ceases to appear as blind utilitarianism or a refusal to ascend toward the divine Idea. The very *activity* of the mathematician,[79] apart from any transformation of the object into a metaphor, provides an explanation for the metonomy, which has taken over from metaphor. Kepler, for his part, continued to think in terms of a resemblance between the universe and a higher order, but he would turn to a musical order to make good the deficiencies of the purely geometric order.[80]

CHAPTER TEN

# The Musical Metaphor

## *The Outcome*

For Kepler, *Harmonices mundis* was the crowning achievement of his life. An unbroken continuity links this book of 1619 to his first essay, *Mysterium cosmographicum* of 1597:

> What I presumed twenty-two years ago, even before having discovered the placement of the five regular polyhedrons among the celestial orbs; what I was profoundly persuaded of before having read Ptolemy's *Harmonies*; what I promised my friends by the announcement of the title of this fifth book, even before I was sure of the thing itself; what I had fixed upon as my subject of research sixteen years ago in a public document; the reason I have spent the best part of my life devoted to astronomical studies, that I went to see Tycho Brahe and chose Prague to live in: this I have finally brought into the light.... (GW, vol. 6, p. 289)

The continuity is most evident from the formulation of what later will be called Kepler's third law, and appears here as eighth among thirteen fundamental astronomical propositions: "But it is absolutely sure and exact that the proportion between the periods of any two planets is precisely related by a ratio of 1.5 to 1 to the proportion of their average distances" (GW, vol. 6, p. 302). As in *Mysterium cosmographicum*, the main consideration is *proportion*. This concern for eurythmy remained a constant for Kepler. As Robert Haas has noted, the little space that Kepler devoted to physical considerations in *Harmonices mundis* is significant in this respect.[1] Kepler was interested in presenting proportions less as the effect of a relation among forces than as the very goal of Creation. The framework is not causal but teleological, and integrates even the ellipse, as we saw in the preceding chapter, into a formist conception of the universe.

*Weaving the Metaphor*

Dating back to Pythagoras, taken up by Plato and Aristotle, treated at length by Boethius, integral to the ideas of the Renaissance starting with Marsilio Ficino, the object of poetic meditations and allegorical representations – the harmony of the universe is one of the most constant themes in the praise of God's handiwork.[2] In the music of the spheres as in other domains, Kepler was an innovator who profoundly transformed traditional beliefs without departing from the received teleological framework. His originality within this framework is multifaceted.

To begin with, Kepler transformed the foundation of the musical theory that forms the basis of the conception of celestial harmony. Starting with Pythagoras, this theory had been based on the ratios among the first four whole numbers. Such a theory "does violence to the natural instinct of hearing" (GW, vol. 6, p. 99): it excludes some melodic intervals and accepts others that are not melodic. Ptolemy, whose treatise on harmony was published as an appendix to *Harmonices mundis*, had already emended the Pythagorean system, but remained under the influence of "abstract numbers" (GW, vol. 6, pp. 99–100). In the sixteenth century, the absence of agreement between theory and practical experience had led Zarlino to add 5 and 6 to the hitherto tetradic theory.[3] Others like Vincenzio Galilei came to mistrust any attempt to reduce harmony to an *a priori* conceptualization.[4] Kepler made a new effort to reconcile the *a priori* and *a posteriori*, speculation and fact. His goal was to uncover "causes that on the one hand satisfy the judgment of the ear, and on the other establish a clear and evident distinction between the Numbers that form musical intervals and those that do not..." (GW, vol. 6, p. 100). Such an effort consisted, here as elsewhere, of shifting from enumerating numbers to enumerated numbers, from an arithmetical to a geometric foundation.

Since harmony is a quality, not a substance, Kepler situated his explanation at the level of *cosmopoetic* figures, that is two-dimensional figures.[5] In conformity with the fundamental significance he attached to the distinction between the curve and the straight line, he linked the problem to the relation between the circle and polygons:

The intersection of the spherical solid by a plane produces a circle, faithful image of the created spirit that governs the body. For the circle is to the sphere as the human spirit is to the divine spirit, that is as the line is to the surface, both being circular. And the circle is to its plane as the curve to the straight line: the relation is incommunicable and incommensurate. With a remarkable effect, moreover, the circle is both in the secant plane that it circumscribes and the spherical solid that it intersects: as the soul is in the body that it informs and to whose form it is linked, depending on God as the body depends on the radiance of the divine visage, and derives from that source a nobler nature. This is why the Circle has been

233

established as the subject of harmonic relations and the source of their terms. (GW, vol. 6, p. 224)

Among the polygons, Kepler isolated those that are "knowable."[6] Knowability (*scibilitas*) is a function of the relation of a polygon to the diameter of the circle it circumscribes: "In geometry, knowing is related to measuring with a determinate unit, which, when figures embedded in a circle are concerned, is the diameter of the circle" (GW, vol. 6, p. 21). The entire first book of *Harmonices mundis* is devoted to the study and classification of polygons in terms of their knowability. The role given to the circle belongs to a properly mathematical tradition, since Euclid, in Book Thirteen of the *Elements* also chose the diameter of the circumscribed circle as the unit of measure for polygons.[7] But logically this choice also belongs to Kepler's geometric symbolism, since the curve continues to assure the intelligibility and the dignity of two-dimensional figures constructed with straight lines.

The movement from geometry to harmony is based on the equivalence, enunciated as early as 1599, between geometric *knowability* and musical *consonance*: "That which is demonstrable in geometry is called consonance in the domain of voice (and motion)" (GW, vol. 14, p. 314). If we imagine the string of a musical instrument as forming a circle, consonances are determined by the arcs subtended by the knowable or demonstrable polygons, that is those that can be constructed with the ruler and compass: the distinction between consonance and dissonance is due to the fact that "the circular line can be divided geometrically in 2, 3, 4, 5, 6, 8, but not 7, 9, 11, 13, not because of a defect in our intelligence or an imperfection in the science of geometry, but by nature" (GW, vol. 16, p. 161). Since the knowable polygons can define an infinite number of arcs whereas the number of har-

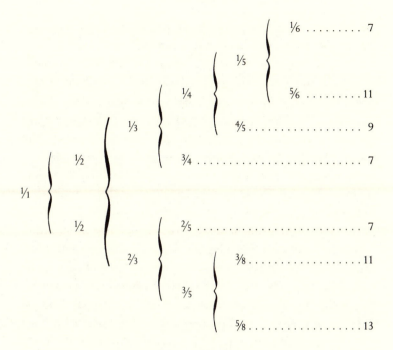

FIGURE 10.1. The derivation of musical consonances according to Kepler.

monic intervals is limited, Kepler added a supplementary principle: the two parts of the circle subtended by one side of a
polygon must be in consonance with each other and with the
total circumference (GW, vol. 6, p. 114). Thus Kepler was able
to obtain all the basic consonances thanks to the subdivision of
a circle by its diameter, a triangle, a square, a pentagon, a hexa-
gon, and an octagon. He brought them together in the table in
Figure 10.1.[8]

The table is obtained by adding the numerator and denominator of a fraction, thereby forming a new denominator to which
are assigned as numerators the numerator and denominator of the

preceding fraction. The process continues until a number appears that characterizes an unknowable polygon (7, 9, 11...).

Starting with a metaphor that compares geometric knowability to musical consonance, Kepler derives a series of associations. The movement is similar to what happens in allegory, which also consists of "weaving" a collection of applications from one semantic field onto another. Developing the metaphor gives "the reader who decodes it a growing impression of correctness"[9] – an impression that is all the stronger if the two fields appear as closed domains bound by a dual application that is systematic and fruitful.

So Kepler succeeded in constructing a complete musical theory, while rigorously checking his deductions against empirical data.[10] The agreement of the two confirmed his initial conviction that music "is not a human invention, subject as such to change, but a construction that is so rational and natural that God the Creator has impressed it upon the relations of the celestial movements" (GW, vol. 6, p. 114). *Harmonices mundis* culminates with an evocation of the musical qualities of those movements.

*The Song of the Planets*
The Renaissance had syncretized multiple ancient conceptions of the harmony of the spheres. Celestial music consisted of a series of interwoven metaphors. Cornelius Agrippa, for example, associated the planets with voices, the strings of a lyre, musical genres, and the Muses. In addition, he associated their distances with musical intervals, and the periods of their revolutions with harmonies.[11] The entire semantic field of music was applied to the heavens. Meshed in several networks of metaphors, each planet was a paradigmatic place referenced by metaphors drawn from different series. Mercury, for example, gave rise to the following associations:

| Allegorical Register | Metaphors |
| --- | --- |
| *Heavens*: voices | *Mercury*: diatessaron |
| strings | hypatehypaton |
| genres | hypophrygian |
| Muses | Calliope |
| distances | semitone[12] |
| period of revolution | double proportion[13] |

Confronting this abundance, Kepler did not distance himself from the system of thinking that was its foundation. Instead, he rethought it in terms of his own musical theory, his heliocentric model of the universe, and the exact numerical data deduced from observation. Celestial harmony is not ordered with reference to the earth, but to the sun. It is not produced by actual distances or motions, but by angular velocities as they would be perceived from the sun: "speculations on harmony consider the eccentric motion of the planets as seen from the sun..." (GW, vol. 6, p. 297). The ratios between the maximum (perihelial) angular velocities and minimum (aphelian) velocities constitute harmonious intervals for almost all the planets. These ratios are said to be "convergent" when they relate the perihelion of the outermost to the aphelion of the innermost, and "divergent" when they relate the perihelion of the innermost to the aphelion of the outermost. All these numerical ratios can be presented concretely, as Figure 10.2 indicates (cf. GW, vol. 6, p. 312).

With the exception of the earth and Venus for the harmonies specific to each planet, and Jupiter and Mars for the harmonies between two planets, the results are very satisfying. The "dissonance" produced by the motions of the earth and Venus, which

237

| Planets | Aphelion (A) Perihelion (P) | Harmonies associated with each planet | Harmonies between two planets | |
|---|---|---|---|---|
| | | | Convergence | Divergence |
| Saturn | A : a P : b | $\dfrac{a}{b} = \dfrac{4}{5}$ (major third) | | |
| | | | $\dfrac{b}{c} = \dfrac{1}{2}$ (octave) | $\dfrac{a}{d} = \dfrac{1}{3}$ (octave + fifth) |
| Jupiter | A : c P : d | $\dfrac{c}{d} = \dfrac{5}{6}$ (minor third) | | |
| | | | $\dfrac{d}{e} = \dfrac{5}{24}$ (2 octaves + minor third) | $\dfrac{c}{f} = \dfrac{1}{8}$ (triple octave) |
| Mars | A : e P : f | $\dfrac{e}{f} = \dfrac{2}{3}$ (fifth) | | |
| | | | $\dfrac{f}{g} = \dfrac{2}{3}$ (fifth) | $\dfrac{e}{h} = \dfrac{5}{12}$ (octave + minor third) |
| Earth | A : g P : h | $\dfrac{g}{h} = \dfrac{115}{16}$ (diatonic semitone) | | |
| | | | $\dfrac{h}{i} = \dfrac{5}{8}$ (minor sixth) | $\dfrac{g}{j} = \dfrac{3}{5}$ (major sixth) |
| Venus | A : p P : j | $\dfrac{i}{j} = \dfrac{24}{25}$ (chromatic semitone) | | |
| | | | $\dfrac{i}{k} = \dfrac{3}{5}$ (major sixth) | $\dfrac{i}{l} = \dfrac{1}{4}$ (2 octaves) |
| Mercury | A : k P : l | $\dfrac{k}{l} = \dfrac{5}{12}$ (octave + major third) | | |

FIGURE 10.2. Harmonic relations between the planets.

exceeds a semitone and is due to the minor eccentricity of their orbits, is of little import. In any case, it in fact cannot be produced, because a planet cannot be simultaneously at its aphelion and its perihelion. As for the relation between Jupiter and Mars, the perfect fit of the pyramid that regulates their spacing — according to the thesis of *Mysterium cosmographicum* (GW, vol. 6, p. 314) – compensates for the harmonic defect.

Encouraged by this success, Kepler continued along the same lines. He demonstrated that the minimum and maximum angular velocities express the distinction between the major mode and minor mode if they are ordered in octaves with the fundamental note corresponding first to the aphelion (major mode), then to the perihelion (minor mode) of Saturn. Similarly, he differentiated the octaves to which the motions of each planet belong and assigned each planet a distinct musical motif (Figure 10.3). The musical notation clearly reveals the different degrees of the eccentricity of the orbits, as well as the greater or lesser distance between the planets: the almost circular orbit of Venus, the accentuated elliptical orbit of Mars, the gap between Mars and Jupiter....

The actual instantiation of the total harmony of the music of the six planets only occurs rarely, "at very long intervals." It is even possible that a perfect musical chord has been produced only once, at the moment of Creation (GW, vol. 6, p. 324). In addition to the harmony of the six planets, Kepler also examined the harmony of five (excluding Venus) and four (excluding Venus and the earth), producing such contrapuntal effects as syncopation and cadence. In the comparison of planets to voices, Saturn and Jupiter are basses, Mars the tenor, the earth and Venus altos, and Mercury the soprano. The *Musica mundana* is polyphonic, and the *musica humana* of the moderns is a replica of it:

FIGURE 10.3. The harmony of the universe according to Kepler. The modern musical notation is by J. L. E. Dreyer, *History of the Planetary Systems from Thales to Kepler* (London: Cambridge University Press, 1906), p. 408.

It is no longer surprising that the art of polyphonic singing, unknown to the ancients, has finally been discovered by man aping his Creator; that, to speak more precisely, man represents the entire duration of the universe in the space of a few minutes with an artificial symphony of several voices; that he tastes something of the satisfaction that God feels for his works, thanks to the sweet impression of delight obtained through music that imitates God. (GW, vol. 6, p. 328)

Kepler sought justification for the smallest details of the musical organization of the universe. He advanced hierarchical reasons to explain that the ratios 1:2 and 1:3 are appropriate for Saturn and Jupiter, the first and highest of the planets, and so on. Such reasoning recalls the *Mysterium cosmographicum*. This time, however, Kepler succeeded where before he had failed: to show perfect agreement between the cosmological hypothesis and the astronomical facts. His third law furnished him with a method for calculating the distance of the planets from the periods of their revolutions, that is from the principle of harmony.[14] The results agreed with Tycho Brahe's observations.[15] If the reality of celestial harmony is thus confirmed, it also explains the inaccuracy of distances calculated on the sole basis of the nesting of polyhedrons.[16] The measurements required by a perfect nesting are not respected, because the higher requirement of harmony requires deviations: the *kosmos harmonikon endekhomenos* takes precedence over the *kosmos geometrikos*.

*Subsequent Research*

Kepler's preoccupation with celestial harmony did not end with *Harmonices mundis*. He took the subject up again in the final sections of the *Epitome*, where he advanced four theses that make even more explicit claims for the perfection of the universe.

The first concerns the relation between the volume of the planets and their distance from the sun. The volume of the planets is modeled on the volume of their orbital spheres: "the ratios among the planetary bodies are identical to those among the radii of their orbits. In terms of the volume, Saturn is thus a little less than ten times larger than the earth, Jupiter more than five times larger, Mars one and a half times; but Venus is a little less than three-quarters the volume of the Earth, Mercury a little more than a third as large."[17]

Another thesis concerns the "abundance of matter" (*copia materiae*) contained in the planets. The proportion of matter is "exactly half the proportion of the volumes or amplitudes (and thus one and a half times the diameters of the spheres and three-fourths the surface areas)" (GW, vol. 7, p. 284). Kepler summarized the relative densities in a table, which compares them to terrestrial materials. As Alexandre Koyré notes, this comes close to reestablishing "the alchemical doctrine, which finds a connection between celestial bodies and metals"[18] (GW, vol. 7, p. 284):

| Saturn | 324 | the hardest gems |
|---|---|---|
| Jupiter | 438 | loadstone |
| Mars | 810 | iron |
| Earth | 1000 | silver |
| Venus | 1175 | lead |
| Mercury | 1605 | mercury |

In addition, the whole of matter is equally distributed among the three parts of the universe: the sun, the mobile universe, and the sphere of fixed stars contain a third each. In a way, we can imagine "that the body of the sun is of gold, the sphere of fixed stars watery, vitreous, or crystalline, and intermediate space full of air" (GW, vol. 7, p. 288).

Finally, the ratio between the sizes of the three principal parts of the universe also obeys mathematical reason: the radius of the sun ($R_1$) is to the radius of the mobile universe ($R_2$) as the latter is to the radius of the sphere of fixed stars ($R_3$): $\frac{R_1}{R_2} = \frac{R_2}{R_3}$ (GW, vol. 7, p. 307).

All these cosmological affirmations are combined with astronomical arguments. Kepler determines the size of the planets, presents a corrected deduction from his third law, calculates the thickness and distance of the sphere of fixed stars. All the elements in the plan of creation have thus been elucidated. And so the mathematical *idea* of creation becomes more and more explicit: a universe without arbitrariness or chance, where every nonidentity participates in proportion, and where proportion, "of all bonds the best," according to the *Timaeus*, "makes itself and the terms it connects a unity in the fullest sense."[19]

## The Place of Man

*Harmonices mundis* crowned Kepler's research in the sense that it provided him with a principle that enabled him to account for the global construction of the universe without repudiating the preceding stages of his thinking. It did not require him to renounce the thesis of the nesting of polyhedrons, and it furnished a teleological explanation for the ellipse. One problem was still unresolved, however: man's relation to this harmony.

Kepler's celestial harmony contains more than an element of ambiguity. On the one hand, this harmony is directed not to the inhabitants of earth but to the sun. Man is effectively dispossessed of any direct perception of the music of the spheres and can only reconstitute it through rational analysis, by mentally adopting a point of view that is not his. On the other hand, this harmony is not even made manifest to the sun except as a kind of *illusion*: it is not produced by true velocities, true distances, but by *angular* appearances. Harmony is established, but not for man, and only as an appearance.... What is the significance then of man's location or the location of the sun?

"Since speculations on harmony envisage the eccentric motion of the planets as they are seen from the sun..." wrote Kepler (GW, vol. 6, p. 309). *As they are seen from the sun....* The idea is repeated insistently throughout Book Five of *Harmonices mundis*. But what does this mean? That man must escape his material situation by means of his *oculus mentis*? That in the final analysis his decentered point of view is not the valid one and that a better perspective exists? If so, then the earth is a poor substitute for one of the foci of the planetary ellipses, the focus that man must try mentally to reintegrate.

Such a situation seems to contradict the thesis, constantly repeated by Kepler, according to which man is the preeminent "contemplative creature" and the earth is the best home that could have been allotted him:

> This globe seems therefore to have been allotted to man with a supreme wisdom, so that he can contemplate all the planets.
>
> It is not in vain that we men of earth can glorify ourselves and must give thanks to God for this very eminent place of habitation given to our bodies. (GW, vol. 4, p. 309)

Even though it no longer occupies the center, the earth is still set apart from the other planets. Its orbit is the third of six. Three celestial bodies (Mars, Jupiter, Saturn) are outside the earth's orbit, and three (Venus, Mercury, the sun) are within; the earth occupies the most suitable location for contemplating this central part of the universe (GW, vol. 4, p. 309).

In addition, Kepler affirmed that the proportions of all the celestial bodies were calculated from the perspective of a terrestrial spectator. While producing his work, the Source (*Destinateur*) of the universe retained as a suitable point of view the very site that would be granted to man. The knowledge of the harmony of the universe that we can acquire from the earth must therefore correspond to the "intended" meaning of the universe:

> In what manner then were the earth's dimensions adapted to the size of the solar globe? — In terms of Vision. For the earth would be home to the contemplative creature, and it was for him that the entire universe had been created. (GW, vol. 7, p. 277)

It is absolutely imperative that man's vantage point not be at the center of the universe:

244

If the earth, our home, did not measure the annual orbit of the other planets — changing from place to place and station to station — human reason would never have arrived at knowledge of the precise intervals of the planets and other things that depend on those intervals; it would never have instituted astronomy. (GW, vol. 6, p. 366)

From his vantage point of the earth, man can arrive through observation and reasoning at knowledge of the plan of the universe. Because the earth is removed from the center and is in motion, and because it completes its orbit at a suitable distance from the other planets and the sun, it is an appropriate home for man, the contemplative creature. It is even possible for man, as Kepler had just proved, to comprehend the harmony of the celestial motions with respect to the sun. A being placed on the sun would see this harmony, but would be incapable of reconstituting through reason the distances of the planets and the exact plan of the universe. The absence of displacement would deprive him of the means to correct by observations made from different points along the earth's orbit the errors that would necessarily slip through our perception because of the inequality of distances along all the planetary ellipses. A solar being could only have an *a priori* knowledge of distances:

> If a mind observes these harmonies from the sun, it is quite obviously deprived of the assistance of motion and the different stations of his place of habitation from which he could tie together the reasoning and arguments needed to measure the intervals of the planets. He compares therefore the daily motion of each planet not as they are in their own orbits, but according to their angular motion with respect to the center of the sun. Consequently, if he knows the size of the spheres,

this knowledge must belong to him *a priori* and cannot be acquired by the labor of reasoning.... (GW, vol. 6, pp. 3–66)

Finally, man's mode of knowing, in contrast to the knowledge of a solar being, reveals the difference between man and God. The relation between earth and sun is a figure for the relation between *ratio* and *mens*. Just as an inhabitant of the sun could only have *innate* knowledge of this harmony, such a creature (if it exists) would be comparable to the Platonic *mens*. Symbolically, it is suitable that man, being a creature of *ratio*, for whom all *a priori* knowledge is hidden "under the veil of potentiality" (GW, vol. 6, p. 226), and whose only access to ideas is by observing visible data, has his place on the moving and decentered earth. His understanding, based on movement "from station to station" around the solar focus, is related to innate knowledge "like the multiple discourse of ratiocination to the simplest mental intellection" (GW, vol. 6, p. 366).

Thus the idea of a universe created *propter nos* is maintained, thanks to the supplementary metaphor of the relation between sun and earth. But it is incontestable that access to the meaning becomes labyrinthine. Kepler's God is like Daedalus.[20]

## Metamorphoses of the Metaphor

### Kepler and Fludd

Robert Fludd's theory of harmony is rooted in the Renaissance. His principal sources are Marsilio Ficino, Cornelius Agrippa, and Zorzi.[21] Frances Yates writes, "At a very late date, after the *Hermetica* has been dated and when the whole Renaissance outlook is on the wane and about to give way before the new trends of the seventeenth century, Fludd completely reconstructs the Renaissance outlook."[22] The late date of Fludd's theory explains why it combines Mannerist and even Baroque traits with the Renaissance legacy. From this perspective, the famous controversy between Fludd and Kepler is extremely interesting.[23] But the open antagonism risks hiding a certain kinship that is no less real: the differences appear all the more clearly, moreover, when we identify their common foundation.

In their approach to the harmony of the universe, both Kepler and Fludd sought to reduce the multiple and variegated symbolism of the Renaissance to a structured system grounded in one symbol. Kepler's harmony, relying on the circle and regular polygons, is ultimately derived from the symbolism of the *sphere* – as are all the forms of Creation. Likewise, Fludd introduced one symbol of origin: the *monochord*, which like the sphere is a figure for the Trinity and informs all of Creation. Peter Ammann writes:

> The only thing distinguishing Fludd's *musica mundana* entirely from the antecedent tradition is his invention of a concept which unites, comprises and systematizes all previous musical analogies, in particular that of the harmony of the spheres, in one vividly descriptive symbol. To Giorgi [Zorzi] the monochord had been one of many possible similes. But to Fludd the monochord, as a concrete instrument, becomes the symbol *par excellence*, or as he puts it: "exactissimum naturae mundanae symbolum et ipsius veritatis typus" [the most exact symbol of the nature of the world and the figure of truth itself].[24]

Like Fludd, Kepler wanted to deductively systematize the shifting and often protean symbolism of the Renaissance.

Another trait common to Kepler and Fludd, and anchored in the Mannerist sensibility, is a tendency toward the esoteric. Fludd loaded the Apollonian ideal of the harmony of the spheres with esoteric elements.[25] Do we not find the same penchant in Kepler? Is not the God who ordered creation by a harmony that is only discovered two thousand years later to be an esoteric God? Was it not the abundance of arcane symbols that restricted the audience of *Harmonices mundi*?[26] Did not Kepler designate his symbolic practice as a new kind of geometric *cabala*?[27] Basically, is not Kepler's symbolism much more esoteric than that of most of his contemporaries – insofar as it requires that all relations be rethought and reconstructed in terms of the opposition between the curve and the straight line in a heliocentric universe?

But the differences are no less important.

With Kepler and Fludd, the crisis of Mannerism is expressed in two opposing ways. For Fludd, the desire to salvage the content of the symbolic heritage of the Renaissance was predominant; this desire, which provoked controversies not only with Kepler, but also with Mersenne and Gassendi,[28] led him to oppose the revitalization of the sciences, where emphasis is on the quantitative approach. He retreated into a vast synthesis of knowledge from the past, which he sought to protect behind veils of traditional esoterica. Kepler, on the other hand, wanted to salvage not the content but the symbolic mode of thinking by grafting it – a unique undertaking – onto the new heliocentric and quantitative conceptualization of the universe. He took on the task of reconstructing the divine language from its foundations up. The strange originality and the solitude of his attempt also produced an esoteric effect. At the dawn of the scientific transformation of the image of the world, Fludd and Kepler embody two complementary attitudes: for one a drawing back and a refusal; for the other an effort to reconcile the symbolic and scientific. It is this oxymoronic effort that makes Kepler both one of the last men of the ancient world and one of the founders of the new.

The implications of these different choices came into play during the controversy between Kepler and Fludd. The imperial mathematician upbraided the Rosicrucian for the mathematical inexactitude of his symbolic representations: they are "hieroglyphs" unrelated to the "diagrams" that reproduce the real relations among things (GW, vol. 6, pp. 386 and 396). In his response, Fludd proclaimed that mathematical precision was irrelevant to his endeavor. He also expressed doubts about the power of mathematics in general. Mathematics is incapable of producing *concrete* knowledge of things. For example, it only produces an abstract knowledge of music; the study of music as a *tangible* phenomenon is above all a question for physics.[29] Likewise, celestial harmony results from the relations of physical objects: light (active principle) and matter (passive principle), outside the realm of their mathematical representations.[30] Mathematics is also of limited help in the comprehension of the spiritual, which cannot be measured or represented by geometric figures.[31]

If we can apprehend neither the perception nor the meaning of reality in mathematical terms, we must not reduce images to the status of the "diagrams" demanded by Kepler. Their function is not to account for the exact quantitative relations between things. "Who does not know the mathematical proportions of the pyramids?" exclaims Fludd.[32] Such knowledge is of no use in discovering the underlying principles of things.[33] Why favor the practice of the *cipher*, over that of the *number*. For Fludd, the function of geometric figures is "formal," not mathematical. The form of an image addresses the imagination, not mathematical reason. Thus the pyramids, illustrating the relation between matter and light (see Figure 7.2) do not reveal relations that can be counted but *bring to light* the "hidden progression of form in matter" or the "ascension from imperfection to perfection, from impurity to purity."[34]

*Other Permutations*

If Fludd's sources belong to the Renaissance, whereas his taste for deductive systemization and for the esoteric links him to Mannerism, he ended up by defining a dominant function for the image, a function that is already Baroque. The validity of "emblems" or "hieroglyphs" does not depend on an exact correspondence with the real, but on their effectiveness. Contrasting the cipher to the number, the emblem to the diagram, Fludd situated the specific force of images in the suggestions they propose to the imagination, considered as a faculty independent from reason.[35]

The insistence on the *practical* quality of the image, on its concrete, tangible appeal to the imagination is also found in Kircher, notably the author of an *Ars magna lucis et umbrae*: shadow and light, pure tangible qualities insofar as they strike the eye and imagination, are fundamental symbols, associated for Kircher as for Fludd with matter and spirit.[36] A lengthy part of the work is devoted to the geometry of conic sections. But this knowledge only serves as preparation. The title says it well: *ars*. All the knowledge, all the technique of the period are put in the service of the production of a theater of marvels directed principally to the imagination.[37]

Kircher also took up the theme of universal harmony, especially in his *Musurgia universalis*. It contains passages on the harmony of sounds and colors, on the correspondence between microcosm and macrocosm, and so on. Here as elsewhere, Kircher favored images that appeal to the imagination, Fludd's "emblems" as opposed to Kepler's "diagrams."[38] Celestial music is not the principal motif. Nor does Kircher's interest in it lead him to attempt the rigorous reduction of the tangible to mathematical relationships. The music of the spheres must not be sought in numerical relationships, but only in the general arrangement of the parts of the universe, with chiefly the preservation of the earth and man in mind. Thus the sun is neither too far away nor too close, so that the earth does not dry up or freeze.[39] Riccioli, who gave a vast survey of the different theories of celestial harmony, agreed with Kircher, for whom harmony in the precise sense of the term is no longer the issue. Instead, it must be understood "in a metaphorical sense and in terms of an analogy."[40] For Riccioli, this metaphoric harmony cannot be what it still is for Kepler: an "absolute," divine metaphor that informs the very being of the universe. It is no more than the shimmer of a particular arrangement of human signs, not the mathematical reality of things.

For Mersenne, although he wrote a *Harmonie universelle*, the theme of the music of the spheres had become altogether secondary. His concern was no longer archetypes. Kepler's ideas on celestial harmony were not "well enough established to explain other than imaginary consonances," he wrote.[41] The fate that befell symbols of origin, whether Kepler's or Fludd's, was typical. Like Fludd, Mersenne compared the Trinity to a monochord, but the comparison appeared as "corollary 2" of "proposition 11" of the *First Book of Consonances*.... Elsewhere he compared the Trinity to a circle, but for reasons, as we have already seen, that are purely oratorical.[42] The issue of a symbolic language no longer arises; it has been replaced with the didactic, with constructions of the imagination acknowledged as such, with science's capacity for rhetorical effect offered, almost as a residue, to the orator. The image is no longer judged in terms of the truth it contains, but in terms of the practical attraction it exercises on the imagination. The attitude is Baroque: "the question is not truth or falsity but utility or detriment."[43]

Such a conception of the *image* leads to a radical distinction between the different functions of the mind. Tesauro, one of the shrewdest Baroque literary critics, called these functions *guidizio*

251

# CHAPTER ELEVEN

# The Lunar *Dream*

The *Somnium seu de astronomia lunari* is a posthumous work by Kepler, published in 1634 by his son Ludovic.[1] The first drafts go back to at least 1610, but the idea of describing celestial phenomena as they would appear from the moon dates back to Kepler's years of study in Tübingen (OO, vol. 8, p. 670). Early versions were circulated in manuscript form. Kepler thought that the *Dream* was known to John Donne while writing his *Conclavium ignatii*.[2] In addition, the presence of easily recognizable autobiographical elements led – still long before its publication – to legal action against Kepler's mother. She was accused of sorcery because the mother in the *Dream* was endowed with magical powers. The trial lasted five bitter years, and she died shortly after her acquittal.

Several ancient works may have suggested to Kepler the idea of writing a dream about the moon. We should keep in mind the principal models: the voyage through the heavens in *Scipio's Dream*, inserted by Cicero in his *Republic* and annotated by Macrobius; Lucian's *True History*, which includes a stop on the moon beginning in Book One; Plutarch's speculations in *De facie in orbe lunae*, an annotated Latin translation of which appeared with the publication of Kepler's *Dream*. In a letter, the astronomer admit-

ted, moreover, that various utopian writings, and especially
Campanella's *City of the Sun*, were not without influence on his
decision to write a fictional work about the moon (GW, vol. 18,
p. 143). Later we will look into the relationship between the
*Dream* and utopian literature. The publication of the *Dream*
also coincided with the beginning of a new vogue in lunar literature. Within the next several years, the chief among these
included Godwin's *Man in the Moon* and Wilkins' *World of the
Moon* (both in 1638), and Cyrano de Bergerac's *Estats et Empires
de la Lune* (1657).[3]

In this context, the *Dream* possesses an originality resulting
not only from the scientific personality of its author, but also
from its breadth of scope. The *Dream* does not simply provide a
fictional occasion for the veiled exposition of a body of knowledge. In a controlled fashion, it engages shifting networks of
meanings, which are transformed each time the reader modifies
his point of view or his interpretive code.

Literature had always constituted a major attraction for Kepler. He made this clear when he spoke about himself in the third
person in a horoscope of 1597 (OO, vol. 5, pp. 476-77):

> This man was born with the destiny of devoting much time
> to difficult things that are repulsive to others. In his child-
> hood, he undertook versifying before the proper age. He
> attempted to write comedies, chose the longest psalms to
> learn by heart.... In poetry, he tried first to write acrostics
> and anagrams.... He then undertook the most difficult of
> diverse lyrical genres; he wrote Pindaric verses, dithyrambs.
> He embraced unusual subjects [such as] the sun's repose, the
> source of rivers, a view of Atlantis through the clouds. He
> delighted in enigmas, searched out the most subtle figures of
> speech; he amused himself with allegories, wove the tiniest

details into them and even teased them by the hair. . . .

And so we should not approach the *Dream* only as an important document in the history of science, but also as evidence of Kepler's ongoing attraction to literature, an attraction that impelled him toward the labyrinthine forms cultivated especially by the Mannerists. In the passage just cited, Kepler's focus is not on the substance of meaning but on the formal production and presentation of meaning – the extreme or affected conditions of its manifestation. A good number of the notes that accompany the text of the *Dream* show a similar focus. On the subject of the notes, we must mention a perhaps unexpected comparison, but one that is unavoidable, and however superficial, that merits being made if only in passing. The principal literary example of a work in which the author's annotations surpass the length of the text itself is Guarini's *Pastor fido*. In the definitive edition of 1602, the pastoral is accompanied by notes that are much longer than the text. A single note may spread over more than two pages in fine print. Taken altogether, the notes propose a dissection, an analysis, a rationalization meant to expose the motivation behind every detail, the meaning and the classical antecedents of every stylistic figure.[4] They reveal a self-engrossed attitude toward writing, the generation and over-determination of meaning that we also find, *mutatis mutandis*, in Kepler.

## The Title

### "Somnium" 1

Reporting the famous episode of the three dreams in his *Vie de Descartes*, Baillet notes a terminological question. At the outset, in the feverish atmosphere so vividly recreated by Jacques Maritain and Georges Poulet,[5] the *Vie* shows us that the mind

255

of the philosopher was "prepared to receive impressions from dreams and visions."[6] Dreams *and* visions: the coordinating conjunction is not so innocent as it might appear today; it does not simply reflect a rhetor anxious to amplify his speech with a series of synonyms. Later the two terms reappear, but in a disjunctive relation: "What is singular to note, is that in doubt as to whether what he had just seen was a dream *or* a vision, he decided while sleeping that it was a dream."

Certainly *somnium* could apply to dreams in general. But in a more technical language, like Baillet's, it designates a kind of dream, in contrast to *vision*. Kepler occasionally distinguished *somnium* from *phantasm* (GW, vol. 14, p. 432) and demonstrated on several occasions his familiarity with treatises on oneiromancy, or divination from dreams. He had read Artemidorus' *Key to Dreams*, Cicero's *De divinatione*, Macrobius' *Commentary on the Dream of Scipio*, various works by Caspar Peucer, and so on.[7] Macrobius, one of the principal authorities on the matter, and to whose *Commentary* Kepler often referred, distinguished "five main types" of dreams:

1. The *somnium* or *oneiros*, the "enigmatic dream" or "dream properly so called" [*somnium proprie vocatur*], which "conceals with strange shapes and veils with ambiguity the true meaning of the information being offered, and requires an interpretation for its understanding";

2. The *vision* (*visio, horama*), which presents persons or things as they will later appear to us in reality;

3. The *oracle* (*oraculum, chrematismos*), "in which a parent, or a pious or revered man, or a priest, or even a god clearly reveals what will or will not transpire, and what action to take or to avoid";

4. The *nightmare* (*insomnium, enypnion*); and

5. The *phantasm* (*visus, phantasma*), which are of no interest

to divination because they are due solely to bodily disorders and contain no revelation.[8]

These distinctions were known and utilized in the sixteenth and seventeenth centuries. Cornelius Agrippa stressed the contrast between the *somnium* on the one hand and the *insomnium* and *phantasm* on the other: "Here by 'dream,' I understand not a phantasm or insomnia, for those are vain things, where divination is not possible, but originating from sleeplessness, fatigue, and bodily disorders...."[9] Caspar Peucer adopted Macrobius' classification in detail.[10]

In this context, the title *Somnium* can have a precise meaning. The correspondence with Macrobius' definition is all the more striking since Kepler himself furnished the allegorical keys for his *Dream*. Macrobius characterized one variety of dream as "universal," when the dreamer "dreams that some change has taken place in the sun, moon, planets, sky, or regions of the earth."[11] The moon is thus placed among the exemplary subjects of *dreams* in the narrow sense of the term.

*"Somnium" 2*
Unlike the dreams of Descartes recounted by Baillet, Kepler's *Dream* does not constitute a biographical or mythographical document. The question of its lived reality simply does not arise. It is a *written text*, for which the appropriate operation is not reconstituting an experience but analyzing the meaning.

The *Dream* is the imitation of a dream, and as such it belongs to a very widespread literary genre. Macrobius classified the literary dream among "fables" that it is appropriate for the philosopher to use. In particular, the genre makes it possible for the philosopher to choose characters "suited to the expression of his doctrines." An example is Scipio the Younger in Cicero's *Republic*. Such a dream "rests on a solid foundation of truth, which is treated in a fictitious style." The content is always allegorical, and "only eminent men of superior intelligence gain a revelation of her [Nature's] truths."[12]

The dream is a literary genre that lends itself to every kind of ambiguity. On the one hand, it is a discourse that appropriates, through imitation, a form of revelation. On the other hand, it only imitates this form. Dreams are, as Macrobius says, "fables – the very word acknowledges their falsity." Revelation and lie, truth and fable, go hand in hand. Kepler thoroughly exploits this superimposition. Duracotus' narrative is a serious and weighty subject, one that can reveal truth since it interposes a veil on a double allegory: the characters represent abstract notions whereas the moon revolving around the earth is a figure for the earth revolving around the sun. But as a fable and lie, the same story also stakes out a field for pleasure: it furnishes the author with an occasion to give free rein to his taste for analogical elaborations – a propensity that he must have had to, if not exclude, at least rigorously control in his scientific work – and it also permits him to twist an inner prohibition into the very heliocentric thesis that the work aims to promote. As a fiction, and on that level alone, it employs elements borrowed from ancient cosmology. These are elements to which the imagination remains attached in spite of everything. In addition, the act of constructing the fable opens a space for the critique of a particular fictional form.

The choice of the dream genre seems to have been made for its positive possibilities and not as the act of prudence that one might imagine, in the defense of theses not acceptable to some authorities. Elsewhere in fact, Kepler confirmed that the veil of fiction does not offer sufficient protection to one who affirms dangerous truths, citing More and Erasmus as examples (GW, vol. 18, p. 143). In addition, such prudence would have contradicted his habitual attitude. In the defense of heliocentrism, he never beat around the bush.[13] Finally, the veil is lifted in the work itself by the notes that accompany the text of the dream.

## The Narrative

### The Narrative Frame

> In the year 1608 there was a heated quarrel between the Emperor Rudolph and his brother, the Archduke Mathias. Their actions universally recalled precedents found in Bohemian history. Stimulated by the widespread public interest, I turned my attention to reading about Bohemia, and I came upon the story of the heroine Libussa, renowned for her skill in magic. It happened one night that after watching the stars and the moon, I went to bed and fell into a very deep sleep. In my sleep I seemed to be reading a book brought from the fair. Its contents were as follows.
>
> My name is Duracotus. My country is Iceland.... (KS, p. 11)

Thus begins the *Dream*. The *narrator* of the frame narrative moves back through time to evoke a night and a dream. In the dream, a book appears to him, and he transcribes the book. He is transformed into a *scribe*, copying the text of another "I" named Duracotus. The text of the book that appears in the dream is accompanied by copious notes (written around 1620–1623). These constitute a commentary that is longer than the narrative; the narrator-scribe takes on the role of *reader*, interlacing the dream text with explanations and interpretations. Among other things, these notes attest to the similarity between Duracotus and the narrator of the frame narrative, the latter examining himself in the former and identifying with him as the principal *actor*.

Thus the *Dream* takes on the appearance of a narrative within a frame. We know the vogue for this form in Renaissance literature, starting with Boccaccio. The *Dream* displays the ambiguity of any narrative within a frame – in addition to the ambiguity inherent in the dream genre. On the one hand, the nesting of the narratives implies a structure consisting of multiple levels, a structure that makes it possible to clearly distinguish various voices and expressive situations. Different perspectives are projected on tiers of narrative material organized according to the principles of a hypotactic narrative. On the other hand, the procedure lends itself to a play of mirrors, such that the general narrator's disappearance is only apparent; he continues to bear a definite relation to the narratives associated with the secondary narrators. Thus Duracotas' narrative reflects back to the general narrator of the *Dream* the image of a double, of another who is the same. The (con)fusion of the two characters is made all the easier by the fact that in their respective narratives both are narrator and actor, and both constantly employ the pronoun "I."

One function of the notes is to reveal precisely this solidarity among levels, drawing attention to the similarities between Duracotus and the general narrator: uncertainties about the separation between the frame and the inner narratives, mutual overspilling from one to the other, substitution of the continuous for the discontinuous. Another function of the notes, as though the frame were never complete, is to serve as a frame for the frame, since they contain the commentary of a more general authority than the general narrator, which they situate, and whose words they clarify.

A reduplication of what is in the frame corresponds, moreover, to the framing of the frame. The general narrator speaks (1) of a *dream* where there appears (2) a *book*. The *Dream* is a book that recounts a dream. But the dream recounts a book. There is a symmetrical inversion of relations: from the book to the dream, from the dream to the book. Here is another indication, if any is needed, that the multiplication of narrative levels only creates an apparent discontinuity: from one level to the next, the signifier stays the same.

But the *Dream* does not stop there. Within Duracotus' book there appears a discourse spoken by a daemon who furnishes scientific clarifications on lunar geography, astronomy, and biology. The daemon's words, at the deepest layer of the dream, are like a rejoinder to the outermost layer constituted by the notes. Two forms of knowledge are grafted onto two narratives. One (in the notes) is to be read in a discontinuous form; the other (in the daemon's words) in a continuous form. On the other hand, the daemon speaks in Icelandic. Kepler finds in him the image of his own knowledge about the moon, but in the mouth of another and in another language – a little as Cervantes discovered his own narrative in *Don Quixote*, but written by another in Arabic. In addition, the same ambiguity exists between Duracotus and the daemon as between the former and the general narrator: one is the double of the other; the daemon explicitly adopts the point of view of "us men."[14]

All the layers can be summarized by Figure 11.1 on the following page.

Duracotus' book, which appears in the dream, and the daemon's monologue, presented in this book, are written – like the notes and the frame narrative – in direct discourse. But the distinction between the words spoken by the different characters offers only a simulacrum of authenticity, denied by the network of correspondences among the levels of the work. At the same time, the contents of the book in the dream and of the monologue in the book call attention to the fiction insofar as they are articulated by invented characters. But the same content is "true"

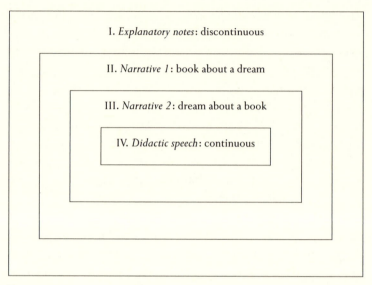

I. *Explanatory notes*: discontinuous

II. *Narrative 1*: book about a dream

III. *Narrative 2*: dream about a book

IV. *Didactic speech*: continuous

FIGURE 11.1.

insofar as it makes explicit, as the notes confirm, a partial model of the author's universe. The Italian critic Mazzoni wrote in 1587, "The idea can be feigned in such a manner that it appears false, or in such a manner that it appears true."[15]

To the similarities of the narrative frames are added descriptive (diegetic) resemblances. Libussa, whose story is recounted in the book read by the general narrator before he falls asleep, is a magician, like Fiolxhilde, Duracotus' mother. Fiolxhilde practices her rituals of invocation at a crossroads, a place sacred to Hecate, the lunar goddess. Duracotus' voyage to Denmark is a first initiation to knowledge of celestial phenomena, which prepares him to understand the daemon's speech. Iceland is a distant and relatively small island compared to Europe – reproducing the relation between the moon (which Kepler calls elsewhere an island) and the earth. Kepler stresses in a note that "Mount Hekla, the Icelandic volcano," is sometimes designated as the location of Purgatory, as was the moon in "pagan theology" (KS, p. 34). We should note, finally, that like the moon Iceland is characterized by disproportion or excess. In Iceland, the days and nights tend to be either very long or very short (a twenty-four hour day occurs around St. John's Day), which gives them a strangeness prefiguring the strangeness of lunar days: "A night and a day, taken together, equal one of our months" (KS, p. 17). On the other hand, the longevity of the inhabitants of Iceland is remarkable: Duracotus' father died after about 70 years of marriage at the age of 150, when Duracotus was only 3. These numbers come back to memory (by counterpoint) when we learn that on the moon the inhabitants have "a short life...but enormous growth, so that nothing would attain a stable state and everything would perish in the midst of its development" (KS, p. 130).

### The Daemon and the Umbra

Duracotus was raised by his mother, Fiolxhilde, who supported them by selling magical herbs, useful to navigators. Driven away by his mother following a careless blunder, Duracotus travels to Denmark and the island of Huen. Warmly welcomed by Tycho Brahe, and enthusiastic about the study of the heavens, he does not return to his boat but spends five years studying astronomy.

Back with his mother, with whom he has been reconciled, Duracotus discovers with surprise that the old woman's knowledge, founded solely on practical experience, is not inferior to the knowledge of the Danish savant. Fiolxhilde tells her son of her relations with the daemons of Levania, the moon. She even decides to initiate him into her art. And so, one spring night under the sign of Taurus when the waxing moon enters into conjunction with Saturn:

[m]y mother went away from me to the nearest crossroads. Raising a shout, she pronounced just a few words in which she couched her request. Having completed the ceremonies, she returned. With the outstretched palm of her right hand she commanded silence, and sat down beside me. Hardly had we covered our heads with our clothing (in accordance with our covenant) when the rasping of an indistinct and unclear voice became audible. (KS, pp. 14–15)

The daemon that appears at that moment gives Duracotus and his mother a detailed description of the moon, first explaining how one gets there by means of the umbra connecting it to the earth.

How do we explain the association between this daemon and the earth's umbra on the moon? Plutarch's *De facie in orbe lunae* would seem to have played a determinative role. Edward Rosen has stressed that the daemon of the *Dream* is a "terrestrial" daemon.[16] It is clearly the form under which the soul of a dead man survives, since it explicitly adopts, as we have said, the point of view of "us men." Now if the daemon is "summoned from the moon" (KS, p. 62), it is because Plutarch wrote that the moon is exactly the place where this kind of daemon lives.[17] Plutarch also designated the earth's umbra as the path by which these daemons move back and forth between the earth and the moon: "Hell is not only the Earth, but also the umbra that accompanies it. That is where souls are found after death. They rise to varying heights depending on how pure they are.... The best can reach the moon. Then they change nature...they become daemons...."[18] Clearly, the important thing to note here is that the status attributed to the earth's umbra presupposes a geocentric cosmos, with the moon located on the boundary of the metaphysical universe.

The *Dream* contains this paradox: while one of its proclaimed goals is to promote the new concept of a heliocentric universe, the fiction that overlays this defense borrows some of its materials from the ancient geocentric cosmology. This constitutes another proof, furnished in person by one of the promoters of the cosmological revolution, that the geocentric universe, scientifically unacceptable as it had become, nonetheless remained more satisfactory in terms of imaginative associations, identifications, and correspondences.[19]

### Sleep and the Sleeper

If the dream in the strict sense constitutes a form of revelation, most theoreticians also took into account psychophysiological factors that could determine its form, based on a typology borrowed from Aristotle, Hippocrates, and Galen.[20] We find traces of this in Kepler.

The theoreticians insisted on the relation between the content of a dream and what the dreamer was doing before falling asleep, his personality, situation, and habits. All this contributes to the substance from which the form of the dream's expression is shaped. Cicero writes, "And the soul stirs up and agitates the residue of these things that took place in waking thoughts."[21] Kepler's *Dream* is presented as the product of the occupations of the day, since it melds into a single narrative the "residue" of the reading of a magician's story and observations through a telescope. Elements originating from a distant past are also grafted on this outline: the lack of a father, the visit with Tycho Brahe, memories of Prague....

The materials brought together in a dream undergo rhetorical transformations, as Kepler noted in a letter: "Dreams are constituted of oratorical figures and tropes."[22] Aristotle had compared the result of these transformations to an image reflected in water.[23] Imitating this dream effect, Kepler can give free rein to his taste for analogy. A note explains, for example, that the ceremony during which the daemon appears is a transposition of what sometimes happened in the author's Prague observatory (OO, vol. 8, pp. 44–45):

> Whenever men or women came together to watch me, first, while they were engaged in conversation, I used to hide myself from them in a nearby corner of the house, which had been chosen for this demonstration. I cut out the daylight, constructed a tiny window out a very small opening, and hung a white sheet on the wall. Having finished these preparations, I called in the spectators. These were my ceremonies, these my rites. Do you want characters too? In capital letters I wrote with chalk on a blackboard what I thought suited the spectators. The shape of the letters was backwards (behold the magical rite), as Hebrew is written. I hung this board with the letters upside down in the open air outside in the sunshine. As a result, what I had written was projected right side up on the white wall within. If a breeze disturbed the board outside, the letters inside wiggled to and fro on the wall in an irregular motion. (KS, p. 57)

Mechanisms like those described in traditional rhetoric explain how proper names appearing in the *Dream* were formed. *Duracotus* came from the memory of proper names of a similar sonority appearing in the history of Scotland, a country facing the Icelandic sea (KS, p. 30). *Fiolxhilde* is the result of combining *Fiolx*, a word that Kepler had noted in several places on a map of Iceland, and *-hilde*, which enters into the composition of numerous feminine names in German.[24] From Scotland to Iceland, from the geography to the inhabitants, from German to Icelandic: the shifts are metonymic.

Aristotle had already drawn attention to the intensification and the transformation that the dreamer's imagination is subject to when presented with nervous stimuli: "men think that it is lightning and thundering, when there are only faint echoes in their ears...and that they are walking through fire and are tremendously hot, when there is only a slight heating about certain parts...."[25] And so, for Kepler, a cloth covering the head of the general narrator is the stimulus for the episode in the dream when Duracotus and Fiolxhilde cover their heads during the magic ceremony (KS, p. 14).

One category of notes supports this first kind of interpretation. The continuous and contiguous elements of the dream are dispersed in the heterogeneity of occupations and impressions. The interpretation that proceeds along these lines leads to an indefinite dissemination of its subject matter.

*The Allegory of Science*
In addition to an interpretation of the dream in terms of the heterogeneous materials that it employs, combines, and transforms, there is an allegorical interpretation in which characters and events become concretizations of abstract concepts.

According to this interpretation, Duracotus represents Science and his mother Ignorance or purely practical experience. The father stands for Reason; he remains anonymous because Ignorance cannot know the identity of Reason. If Duracotus only begins to write after his mother's death, it is because Science can only appear when Ignorance has been dissipated (KS, p. 36). The daemons embody particular sciences; their name comes from joining the noun *daimon* and the verb *daiein*, "to know"; the specific daemon that appears to Duracotus and Fiolxhilde represents astronomy.[26] The passage through the *umbra*, of Pythagorean origin, takes on a new sense: "If we continue the allegory, [we must

267

understand that] it is easy for reason with the help of shadow measurement [i.e., measurement of the *umbra*] to attain knowledge of celestial phenomena" (KS, p. 63).

Such an oneiromantic interpretation, substituting an abstract text for the narrative, is completely classical.[27] To be fully achieved, this kind of reading must replace the fiction with another discourse that sticks to its own field of interest, while coherently utilizing all the components of the stylized structure. The allegorical meaning shows itself to be the transcendent support, the origin and the exclusive end of the letter of the text. In the commentary on the *Dream*, however, things do not quite work out this way. The allegorical sense does not function as origin and end of the fiction, as something on which the fiction depends entirely, but which could exist without it.

Thus some notes express a lack of certainty about which allegorical key to adopt — a lack of certainty by which the writer distances himself from the hidden meaning. The daemon of astronomy is called by a name composed of twenty-one characters. Kepler claims he is perplexed, because this could correspond to the number of letters in *Astronomia copernicana* or the number of possible conjunctions between the planets, or even the number of possible combinations on a throw of the dice....[28]

The allegorical reading, moreover, is not the only possible transformation of the letter of the *Dream*. We have seen that the text can also be understood as tying together materials originating from the author's personal past, from legends, and so on. This other reading leads to explanation rather than interpretation. Duracotus and Fiolxhilde can be *interpreted* as incarnations of universals, but also *explained* as products of the imagination, which engenders their names, their personalities, their actions, according to a particular rhetoric that is not mastered. Explanation scatters the narrative units, makes them heterogeneous, whereas interpretation unifies them, or tends to unify them, to make them homogeneous. This dissociation is worth emphasizing. The two readings are not aligned on an axis at the same textual depth but correspond to signifying configurations linked to the adoption of different points of view. Explanation and interpretation are not superimposed. Their articulation is not even made explicit. They are simply juxtaposed.[29]

### The Non-Rationalized Imagination

Beyond the explicitly proposed allegory, the narrative of the *Dream* is rooted in the images and structures that anthropology finds in the mythologizing imagination.

Thus an entire network of narrative and descriptive elements corresponds to the projection of mythical associations: mother, island, water, herbs are inscribed in the framework of traditional lunar themes.[30]

In addition, Duracotus appears, especially at the beginning, as a kind of lunar Prometheus. Bachelard notes that "the problem of obtaining a personal knowledge of fire is the problem of *clever disobedience*. The child wishes to do what his father does, but far away from his father's presence, and so like a little Prometheus he steals some matches."[31] Substitute the young Duracotus for the child and the lunar figures of the mother and the sachet of stolen herbs for the solar figures of the father and fire, and we have a similar situation. Thus Duracotus becomes and remains the symbol of what Bachelard calls the *will to intellectuality*.[32]

As son, Duracotus also represents continuity and progression, becoming. He becomes a figure of mediation between Reason (his father) and Empirical Practice (his mother). But to complete the process, he will have to be engendered a second time. In effect, the *Dream* introduces a second paternal figure, Tycho Brahe, who welcomes, nourishes, and instructs the young Duracotus. In general terms, Gilbert Durand stresses the importance of this motif in the nocturnal (lunar) operations of the image: "there exists in myth a parental doubling of the real father by the mythic father, one of humble origin, the other divine and noble, one the 'false father' or foster-father, the other the true father."[33] Tycho Brahe, thanks to whom Duracotus is born to true science, will also permit him to communicate with his mother after his return to Iceland. Mother and son exchange their respective knowledge, as members of a spiritual community. The son is then ready to receive instruction from another intercessor, the daemon, whose traditional function is to serve as mediator between earth and heaven[34] and whose monologue in the *Dream* effectively represents the communication of knowledge that is generally withheld from men.

### Description of the Moon

Marjorie Hope Nicolson and subsequently others have seen in Kepler's *Dream* the source and origin – *fons et origo* – of the modern cosmic voyage and science fiction.[35] To this end, they have emphasized the role of scientific extrapolation. On the one hand, the description of the moon is based on the most up-to-date scientific knowledge of the period, as attest numerous references to the works of Galileo and Kepler himself. On the other hand, starting with this knowledge, the author proceeds by analogy, for "in a dream it is necessary to have the freedom sometimes to invent even that which was never perceived" (KS, p. 89). Depiction of life on the moon, for example, is based on terrestrial models, and tries at least to be plausible: since there are "immense alternations of heat and cold coming directly on each other's heels," Kepler extrapolates from works on Africa and the Nordic regions to create his picture of lunar life (KS, pp. 130–31).

But beside the extrapolation (obvious and often annotated), there are two other important characteristics that the *Dream* shares with science fiction: the suggestion of distant knowledge and a utopian connection.[36]

### The Inverted Universe

The daemon's description of the moon is not limited to meaning what it says. Several notes affirm that Kepler intended to develop, through the example of the moon, "an argument in favor of the motion of the earth or rather a refutation of the arguments, based on sense perception, against the motion of the earth" (KS, p. 82).

To this end, Kepler employs the technique of inverting the representation of the universe. The lunocentrism of lunar creatures corresponds to the geocentrism of earth-dwellers: "For Levania seems to its inhabitants to remain just as motionless among the moving stars as does our earth to us humans" (KS, p. 117). A note elaborates:

Here you have the principal thesis emphasized in a full statement. Of course, we earth-dwellers think that the plane on which we stand, and together with it the balls on our towers, remain stationary, but around those balls the heavenly bodies revolve in their travels from east to west.... For in like manner the moon-dwellers, too, believe that their lunar plane and the ball of Volva hanging up high over it remain in one place, although we know for a certainty that the moon is one of the movable heavenly bodies. (KS, p. 101)

The inhabitants of the moon worship the earth just as earth-dwellers worship the moon:

But the most beautiful of all the sights on Levania is the view
of its Volva. This they enjoy to make up for our moon.... 
(KS, p. 21)

The image of the inverted universe is developed in detail through
the exposition of astronomical calculations based on the belief that
the moon is stationary and occupies the center of the universe.

Just as the observation of celestial phenomena permits humans
to divide the earth into five zones, it also permits the inhabitants of Levania to perceive two hemispheres on their "planet";
one, *Subvolva*, is always pointed toward *Volva*, the earth or "the
sphere rotating around its own motionless axis" (KS, p. 79); the
other, *Privolva*, is deprived of that view. The circle that separates
the two hemispheres is called the *divisor*.

In Levania, the days are always about equal to the nights. For
the Privolvans, day is a little shorter than the night, whereas the
opposite is true for the Subvolvans. Together, day and night make
up one of our months. The sun advances through the zodiac at
the rate of one sign per day. The common year is nineteen of our
years. During this year, the sun rises 235 times, the fixed stars
254 times. For the Subvolvans, the sun rises when the moon
appears to us in its fourth quarter; for the Privolvans, when it is
in the first quarter.

The inhabitants of the moon have their summer and winter,
but seasonal variations are smaller and do not always affect the
same places at the same time. In addition, the torrid and frigid
zones are proportionally much less extensive than on earth.

Planetary motions are much more complicated as seen from the
moon. Beyond the irregularities that we observe, there are three
others, two longitudinal, one latitudinal; the first has a period
of 1 lunar day, the second of half years, the third of 19 years.

The earth waxes and wanes, analogous to the waxing and

waning of the moon as perceived from the earth. These phases correspond to the four parts of the lunar day, which a more precise observation of the displacement of the spots of Volva lets us further subdivide.

A total eclipse of the sun on the earth corresponds to a partial eclipse of the earth on the moon. An eclipse of the moon on the earth is an eclipse of the sun on the moon. The Subvolvans see all eclipses, whereas on the earth half escape our view and can be seen only from the opposing hemisphere.

There is no need to continue adding details. Their sense is clear: a living being on the moon could elaborate an astronomy and calculate time *by taking the moon for the stationary center of the universe*. Obviously, this demonstration could be replicated for Venus, Jupiter, or Saturn. Each system would be relative to its point of view, but none would correspond to reality. Already when studying at Tübingen, Kepler had proposed a similar demonstration on the subject of the moon.[37] In the *Astronomia nova*, he transferred the position of the observer to Mars, to study the earth's motion from that perspective (GW, vol. 3, p. 22). In the preface to the *Dioptrics*, he imagined the earth as seen from the moon or Jupiter. Galileo also recommended the strategy of adopting a lunar point of view.[38]

The lesson that can be drawn is already Cyrano's lesson: "The moon is a world like this one, to which ours serves as a moon." The traditional argument − also recalled by Cyrano: "what great likelihood is there for you to imagine that the sun is stationary, when we see it move? and what appearance that the earth turns so rapidly, when we feel it firm beneath us?"[39] − is shattered by the transposition of moon and earth.

This problem of *appearance* led Kepler and others to engage in a systematic inversion as a way of exposing the lack of *being*. The problem affected both social life and art forms as well as the scientific revolution — we can cite Gracian, for example. In the *Criticón*, a servant climbs a height from which he turns his back on the palace of Falimonde and brings out a mirror, saying: "all things of this world must be seen in reverse to be seen in their truth."[40] The *Dream* is such a mirror, reflecting a reverse image of a (false) earthbound science.

### *Utopia and the* Dream

In a letter to Matthias Bernegger dated December 4, 1623, Kepler mentioned various utopias in connection with the possible drafting of a work on the moon:

> An experiment with the telescope that I carried out recently produced a marvelous sight, altogether remarkable: cities and walls, which were circular because of the shape of the *umbra*. What more should I say? Campanella wrote his *City of the Sun*. And if we were to write a *City of the Moon*? Wouldn't it be excellent to paint the cyclopean mores of our times in lively colors, but leave the earth behind and go to the moon, for the sake of prudence? But what is the good of such evasive action, since neither More in his *Utopia* nor Erasmus in his *Praise of Folly* were so well protected that they didn't have to defend themselves? We must forsake this political tar pit and stay within the green and pleasant plains of philosophy. (GW, vol. 18, p. 143)

The letter emphasizes the satirical aspect of utopian literature. It seems, moreover, to embrace More and Campanella's utopias, as well as Erasmus' satire, in terms of a more comprehensive genre. Utopian literature always contains a satirical aim. To the degree that the utopian text dramatizes a "no-place" (*ou-topos*), its only "positive" referent is the one it creates. But as a picture or representation, it also has a kind of "negative referent in the real society [where] it seeks to provoke a critical consciousness."[41]

In the letter just cited, Kepler mentions utopian literature only to set it aside. The *Dream* nonetheless occupies a precise relationship to this genre, insofar as Duracotus' book constitutes an anti-utopia. The realistic description of the lunar countryside, based on telescopic observations, forms a contrast to possible utopian idealizations. The only "positive" referent pertains to the activity of contemporary science. What normally would be the utopian "positive" referent becomes in this text a "negative" referent. The new positive science negates a positive fiction and does so with the methods of fictional discourse.

On the literary level, Kepler's *Dream* is equivalent to the corruptible and spotted moon observed by Galileo, and painted by Cigoli at the feet of the Virgin.[42] The moon is not a meaningful model of somewhere else. It represents a defective here. Lunar matter is neither immutable nor eternal; and the form of the moon is not the ideal circle, but irregular. Privolva, the side of the moon that is turned away, constitutes a significant element. It does not participate in the purer space of the "metaphysical" universe, but is more desolate and desertlike than Subvolva:

> The whole of Levania does not exceed fourteen hundred German miles in circumference, that is, only a quarter of our earth. Nevertheless, it has very high mountains as well as very deep and wide valleys; to this extent it is much less of a perfect sphere than our earth is. Yet it is all porous and, so to speak, perforated with caves and grottoes everywhere, especially in the Privolvan region; these recesses are the inhabitants' principal protection from heat and cold. (KS, p. 27)

The invented part of the description, concerning the forms of lunar life, pursues this anti-utopian purpose. On the moon, life is extremely primitive and all terrestrial proportions are distorted:

Whatever is born on the land or moves about on the land attains a monstrous size. Growth is very rapid. Everything has a short life, since it develops such an immensely massive body. The Privolvans have no fixed abode, no established domicile. In the course of one of their days they roam in crowds over their whole sphere, each according to his own nature: some use their legs, which far surpass those of our camels; some resort to wings; and some follow the receding water in boats; or if a delay of several more days is necessary, then they crawl into caves. Most of them are divers; all of them, since they live naturally, draw their breath very slowly; hence under water they stay down on the bottom, helping nature with art. For in those very deep layers of the water, they say, the cold persists while the waves on top are heated up by the sun; whatever clings to the surface is boiled out by the sun at noon, and becomes food for the advancing hordes of wandering inhabitants. For in general the Subvolvan hemisphere is comparable to our cantons, towns, and gardens; the Privolvan, to our open country, forests, and deserts. Those for whom breathing is more essential introduce the hot water into the caves through a narrow channel in order that it may flow a long time to reach the interior and gradually cool off. There they shut themselves up for the greater part of the day, using the water for drink; when evening comes, they go out looking for food. In plants, the rind; in animals, the skin, or whatever replaces it, takes up the major portion of their bodily mass; it is spongy and porous. If anything is exposed during the day, it becomes hard on top and scorched; when evening comes, its husk drops off. Things born in the ground — they are sparse on the ridges of the mountains — generally begin and end their lives on the same day, with new generations springing up daily.

In general, the serpentine nature is predominant. For in a wonderful manner they expose themselves to the sun at noon as if for pleasure; yet they do so nowhere but behind the mouths of the caves to make sure that they may retreat safely and swiftly.

To certain of them the breath they exhaust and the life they lose on account of the heat of the day return at night; the pattern is the opposite of that governing flies among us. Scattered everywhere on the ground are objects having the shape of pine cones. Their shells are roasted during the day. In the evening when, so to speak, they disclose their secrets, they beget living creatures. (KS, pp. 27-28)

The presentation of a utopia generally takes the form of a (travel) narrative, which makes description possible. This description represents the "specialized mode of discourse that is appropriate for utopia." The linear progression of a story is intended to lead to the unfolding of an exemplary situation. The genre demands a reading that goes "from the whole to the part, from the whole to the detail, a kind of reading that is precisely an iconic reading based on the coexistence of all its elements."[43] In utopian texts, the narrative serves especially to introduce, to prepare and justify the description in which it will eventually be absorbed. The fictional description finds its motivation in the fictional narrative, which makes it plausible as a fiction. This is the kind of relationship that we encounter in the *Dream*: the narrative literally stops the moment the daemon appears. The characters fade away in favor of the description of the moon. And as if to emphasize a certain status for the narrative, having become auxiliary to the description, the *Dream* ends abruptly, as the dreamer wakes up: "When I had reached this point in my dream, a wind arose with the rattle of rain, disturbing my sleep…" (KS, p. 28). Thus the text is brought to closure by a fortuitous event, at the level of the frame narrative, but not at all at the level of the narrative within the frame: the reader learns nothing more about its protagonists. Truly concluded in a non-arbitrary fashion, there is only the description of the moon. And as for utopia, the *Dream* not only criticizes an idealization of somewhere else: it also brings to light, by restoring their arbitrariness, the narrative "threads" that govern the genre.

## The Exemplary Nature of the Dream

### Literary Anamorphosis

The notes that accompany the text of the *Dream* render a pluralistic reading. If the narrative contains a scientific allegory, reading it this way does not shift the focus to something like Rabelais' *substantifique moelle* [substantitive marrow]. The allegory is maintained only for the first part of the text and corresponds more to an oblique view of the dream than to its natural depth: the interpretation is sometimes ironically hesitant and the generation of the text as dream is not at all presented as the action of placing a veil on a pre-existing allegorical meaning, but rather as a process of transformation, starting from heterogeneous memories and sensations, to which one can *also* apply an allegorical perspective. The daemon's scientific exposition is acceptable if he is understood from the point of view of a lunary creature, but from another point of view this exposition *also* presents the inverse image of terrestrial science. Moreover, the absorption of the narrative into description and the contents of this description *also* constitute a critique of the utopian genre.

Faced with this diversity, we would do well not to speak of superimposed layers of meaning. The various interpretations do not proceed from a fixed point of view to explore successive levels of the text. Each interpretation implies a different perspective, resulting from a continual shift in the point of view applied to the same textual surface. In sum, the *Dream* confronts the reader as a literary anamorphosis. More precisely, it recalls the multiple anamorphoses that Niceron alludes to in his treatise: "And in such a manner one can make six, seven, eight different portraits that will appear to someone who approaches little by little to rise one after another in the mirror and disappear out the top."[44] Each of these representations corresponds to a different combination of marks inscribed on the same underlying structure (the same page); only the motion of scanning the page (the progression of the reading) sets all these appearances in motion.

*The Final Oxymoron*

Dream, book, conversation with a supernatural being: these three modes of presentation correspond to the first three types of what André Festugière, in his study of Hermes Trismegistus, calls the "literary fictions of the logos of revelation."[45] And so Kepler proliferates, while embedding one within another, the narrative forms associated with traditional initiation narratives. Hermeticism, magic, and demonology constitute the frames in which he presents his defense of the Copernican revolution. We must add the planetary *umbra*, whose role is inspired by the geocentric cosmos, as are all the elements derived from the mythological imagination, including the decision to structure the narrative as successive efforts of mediation. The new world appears in a context still composed of the forms of ancient belief. And so the *Dream* reflects in a fundamental way Kepler's complete personality: his laws, co-constituents of a new representation of the universe, appear in a work whose general mind-set is still rooted in the Renaissance. Like Kepler himself, who has often been called

# Conclusion

Copernicus' poetics involves a number of demands that appear, when we first consider them, to be in addition to the demand for truth. But these supplements are ambiguous. Thomas Kuhn notes:

> Copernicus' system is neither simpler nor more accurate than Ptolemy's. And the methods that Copernicus employed in constructing it seem just as little likely as the methods of Ptolemy to produce a single consistent solution to the problem of the planets.... Even Copernicus could not derive from his hypothesis a single and unique combination of interlocking circles, and his successors did not do so.[1]

If the criterion of truth, in the sense of a more complete correspondence with the facts, could not be decisive, it is clear that supplemental requirements involve more than the simple addition of secondary values. On the contrary, they inform the conception of truth and are part of the global scheme in terms of which the truth is presupposed and given a concrete representation.

Among these elements, we have acknowledged, at the very heart of the Copernican undertaking, the conviction that man can know the world order in its reality and totality. The Copernican

281

revolution is based on the idea of an alliance between God and man, an idea characteristic of the Neoplatonism of the Renaissance. The Platonic ideal of a world created *propter bonitatem* – for the *idea* of goodness, of which man only knows the shadow – is replaced with the ideal of a world created *propter hominem*. The fact that man has been expelled from the center of the universe in no way impedes faith in this alliance. *De revolutionibus* never speaks of this as a humiliation, and later Kepler never stopped praising the decentering of the earth: its orbit was for him the best possible vantage point for viewing the universe.

Postulating the *vertical* significance of the cosmos, the transcendence of its meaning, and conferring *anagogical* significance to the discovery of its real nature, Copernicus' poetics is also oriented toward *organicism* and *formism* as presupposed schemas of the natural order to be discovered. On the one hand, the universe is not an assemblage of parts that can be described from alternative points of view. It necessarily constitutes a specific whole: the parts have no relevant characteristics independent of each other and that whole. And, on the other hand, the qualities of the constituents of the universe are not explained by the context that they provide for each other, but obey a higher final end. Organicism manifests itself concretely in the demand for "symmetry," formism in the thesis of the universal propensity for circular form.

Reliance on these categories and schemas, as well as the concrete content associated with them, is tightly linked to the intertextual field of the Renaissance. The presuppositions that inform Copernicus' procedures, the shape of the solutions he proposes, are all related to ideas defended and responses adopted elsewhere – in philology, the theory of art, and so on. Even the way in which he conceives his place in the history of astronomy (the need for a *renovatio*, autonomy from theology) is a function of the cultural context of the Renaissance. A poetics is obliged to seize the scientific hypothesis at the precise moment it emerges from its cultural milieu, but while it is still in the mixing pot of texts, prejudices, symbols, and so on. We could never hope to understand Copernicus by detaching him from this synchronic field and inserting him only in the diachronic history of astronomy.

I have described the formulation of the Copernican hypothesis, with its principal ramifications, in terms of tropes. At the outset, Copernicus manifested the desire to reform – in terms of enumerated categories and schemas – *mathematical astronomy*. This desire leads to the transposition, by way of *metaphor*, of the requirement for "symmetry," already widespread in Renaissance art. Since a "symmetrical" representation of the universe is only possible with a central and stationary sun, various commonly accepted ideas and beliefs must be replaced. The replacement takes the form of *metonomy* when it is necessary to reconceive the *empirical data*; it corresponds to *synecdoche* when attention turns to the *theoretical elements*. Thus a metonymic operation makes it possible for the empirical perception of the sun to be interpreted as an effect replacing its cause (the earth's motion). In the physical theory, on the other hand, a synecdoche of a part for the whole comes into play when Copernicus replaces the totality of the universe with celestial bodies taken individually, in order to lay a foundation for a conception of natural locations that is appropriate for heliocentrism. The elimination of the theoretical element represented by equants brings into play another synecdochic relation, this time of genus and species: in order to increase the organistic unity, the circles that account for uniform speed are brought into the same genus (having the same "specific" center) as circles representing the orbits. The new image of the universe produces in turn a new tropological *effect* through the possibility of associating the solar center with the conception of the divine. If focusing on this theme accentuates the insertion of *De revolutionibus* in the total textual field of the Renaissance, it also enhanced for some (like Kepler) the attractiveness of the theory of which it is the effect.

Examining the poetic presuppositions of Copernicus, Kepler makes explicit and enlarges the scope of some new potentialities. The theme of the solar center leads to the allegory of center, radius, and surface as a figure for the Trinity. Since the symbol of the Father occupies the center, it is conceived as the source of the energy that maintains the planets in their orbits: favoring the center is no longer an effect, as with Copernicus, but a stimulus. On the other hand, starting with the *Mysterium cosmographicum*, formistic thought gains new strength by contrasting the curve and straight line as figures for the Creator and Creation. Henceforth, rectilinear motion and rectilinear figures also acquire dignity in the supralunary cosmological order. Next, another figurative network, of a musical origin, is superimposed on the geometric figures and explains, through its superiority, why geometric figures are imperfectly realized in nature. The concept of harmony justifies the replacement of the circle by the ellipse and explains the gaps in interplanetary space with respect to the regular polyhedrons. Thus, the world order is *both* geometric and musical. Organistic thinking can be maintained only in terms of a hierarchy that discounts geometric requirements when they conflict with harmonic requirements.

The same double interpretation and hierarchy also appear in the combination of contextualism and formism. On the one hand, the motion of the planets is explained by their contextualist relation to a *species* emanating from the sun. But on the other hand, the resulting elliptical form itself makes possible harmonic intervals. Contextualism is subordinate to a formistic final cause. Thus we find the double system of causality that Gérard Simon already noted in Kepler: the one, which we call formism, implies "a causality of the analogical-hierarchical type, one could almost say of participation" and reveals musical phenomena in the elliptical orbits. The other, called here contextualist, "transitive direct causality," implies a *contact* (a material relation, the *species*).[2] Certainly, it is this type of causality that produces Kepler's (relative) "modernity." But from our point of view, it is significant that contextualism seems to be "topped off" with a formism that never loses sight of the meaning of phenomena in a divine *plan*. Kepler's poetics seeks neither more nor less than to reveal that plan.

This poetics remains anchored in a philosophy of the *Idea*, of the divine *disegno interno*. If it is in Mannerism that we find the most systematic attempt to reelaborate a theory of ideas that could serve as a basis for an aesthetics, this attempt finds its exact match in Kepler's theory. Kepler systematically reelaborated – taking into account all the problems of the new image of the world – a theory of the divine Idea as expressed in the *disegno esterno* of a world whose form appears more and more problematic and hermetic, multiplying obstacles and refinements for the individual who is attempting not only to describe, but also and especially to interpret.

Kepler's poetics finally gives rise to an immense *hypercodification* of the world. If Copernicus attempted to construct a theory that would meet the principal demand for "symmetry," Kepler multiplied the supplemental motivations: the relation between the sun, planets, and the fixed stars, the distance and appearance of the fixed stars, the number and order of the planets, the relation between their respective motions as well as their volumes, the distribution of matter among the three great regions of the universe – all these take on meaning with the introduction of a multiplicity of codes grafted onto the global order. The taste for "collecting" – in this case codes – does not disappear from this "new astronomy," even if it seeks also to be based entirely on "physical" causes. Kepler's universe, entirely subject to "physical" laws, is also an immense collection of "metaphysical" curiosities, the preeminent *Wunderkammer* or *Kunstkammer*. The fact that Kepler wants his work to be scrupulously *exact* at the same time, in accord with observed facts, never sacrificing one requirement for another, gives it the *oxymoronic* allure that characterizes it in depth.

# Notes

ABBREVIATIONS

OR    Nicholas Copernicus, *On the Revolutions*, J. Dobrzycki (ed.), trans. E. Rosen (Baltimore: Johns Hopkins Press, 1978).

3CT    *Three Copernican Treatises*: *The Commentariolus of Copernicus*; *The Letter against Werner*; *The Narratio prima of Rheticus*, trans. E. Rosen (New York: Columbia University Press, 1939).

OO    *Johannis Kepleris Opera Omnia* (*The Complete Works of Johannes Kepler*), C. Frisch (ed.), (Frankfurt-Erlangen: Heyden and Zimmer, 1858 *et seq.*)

GW    Kepler, J., *Gesammelte Werke* (*Selected Works*), W. Von Dyck, M. Caspar, *et al.* (eds.), (Munich: Beck, 1938 *et seq.*).

KS    *Kepler's Somnium: The Dream or Posthumous Work on Lunar Astronomy*, trans. E. Rosen (Madison: University of Wisconsin Press, 1967).

INTRODUCTION

1. Charles S. Peirce, *Collected Papers* (Cambridge, MA: Harvard University Press, 1931), vol. 5, p. 590. On Peirce's notion of *abduction*, see, among other publications, the special issue of *VS* 34 (Jan.–April 1983), entitled *Abduzione*. On the relations between *abduction*, *deduction*, and *induction*, see especially *Collected Papers*, vol. 2, pp. 623–44; vol. 5, p. 590; vol. 6, pp. 469–75; vol. 7, pp. 202–09. For Peirce, *abduction* ranges over the entire realm of the possible, whereas *deduction* depends on the necessary (stating the consequences that flow

from the acceptance of a hypothesis) and *induction* on the real (inferring agreement or disagreement between reality and a hypothesis or a theory).

2. *Ibid.*, vol. 5, pp. 173, 591, 603.

3. Carl G. Hempel, *Philosophy of Natural Science* (Englewood Cliffs, NJ: Prentice-Hall, 1966), ch. 2.3.

4. Karl Popper, *Conjectures and Refutations* (London: Routledge & Kegan Paul, 1972), p. 192.

5. Thomas S. Kuhn, *The Structure of Scientific Revolutions*, 2nd ed. (Chicago: University of Chicago Press, 1970).

6. Thomas S. Kuhn, *The Essential Tension* (Chicago: University of Chicago Press, 1977).

7. Joseph D. Sneed proposes a reformulation of Kuhn's theory with the help of set theory. See Sneed, *The Logical Structure of Mathematical Physics* (Dordrecht, Holland: Reidel, 1971).

8. Kuhn, *The Structure of Scientific Revolutions*, pp. 122–23.

9. Stephen Toulmin, *Human Understanding* (Princeton, NJ: Princeton University Press, 1972), vol. 1.

10. Wladyslaw Krajewski, *Correspondence Principle and Growth of Science* (Dordrecht, Holland: Reidel, 1977).

11. Dudley Shapere, "The Character of Scientific Change," in T. Nicles (ed.), *Scientific Discovery, Logic, and Rationality* (Dordrecht, Holland: Reidel, 1980), p. 68.

12. See, among others, Norwood R. Hanson, *Patterns of Discovery* (Cambridge, MA: Cambridge University Press, 1958).

13. Cf. Jaakko Hintikka, "The Semantics of Questions and the Questions of Semantics," *Acta Philosophica Fennica* 28.4 (1976); and "On the Logic of an Interrogative Model of Scientific Inquiry," *Synthèse* (1981), pp. 69–83.

14. See the references below, where I deal with the question more closely. Michel Meyer unifies the logic of questioning and the logic of metaphor in *Découverte et justification en science* (Paris: Klincksieck, 1979).

15. Lakatos distinguishes "progressive" and "degenerating" research programs. See, for example, Imre Lakatos and Elie Zohar, "Why Did Copernicus's

Research Program Supersede Ptolemy's," in R. S. Westman (ed.), *The Copernican Achievement* (Berkeley: University of California Press, 1975), pp. 354–83. Larry Laudan, who advances the notion of a "research tradition," illustrates well the complexity of the factors involved. See Laudan, *Progress and its Problems* (Berkeley: University of California Press, 1977); and *Science and Hypothesis* (Dordrecht, Holland: Reidel, 1981).

16. Gaston Bachelard, *Psychoanalysis of Fire*, trans. A. C. M. Ross (Boston: Beacon Press, 1964), pp. 5–6.

17. Gerald Holton, *The Scientific Imagination* (New York: Cambridge University Press, 1978). Holton's research is in the tradition established by Bachelard, even though he does not cite Bachelard. In both cases, the study is directed toward presupposed elements: the "false weight of unquestioned values" for Bachelard (*Psychoanalysis of Fire*, pp. 5–6), "presuppositions" or "thematic presuppositions" for Holton (pp. 399 and 426). While restricting himself to the thematics of the four elements, Bachelard notes that "It would be more difficult but also more fruitful to use psychoanalysis to examine the bases for certain other more rational, less immediate and hence less affective concepts than those attached to our experiences of substances." He imagines tasks for the researchers who will eventually follow: "If we succeed in inspiring any imitators, we would urge them to study, from the same point of view as a psychoanalysis of objective knowledge, the notions of totality, of system, element, evolution and development...." Certainly Holton does not confine himself to what Bachelard calls a "psychoanalysis of objective knowledge." Nonetheless Holton's "themata" correspond closely to the items that Bachelard proposed his emulators should investigate: antithetical pairs such as volition-involition, reductionism-holism, complexity-simplicity, and so on, provide "more reasoned, less immediate and therefore less emotional" data than fire, air, water, and earth. See also Holton, *L'Invention scientifique* (Paris: Presses Universitaires de France, 1982).

18. Holton, *L'Invention scientifique*, p. 15.

19. Alexandre Koyré, *Études d'histoire de la pensée scientifique* (Paris: Gallimard, 1973), pp. 11–12.

20. Robert Lenoble, *Histoire de l'idée de nature* (Paris: Albin Michel, 1969), p. 23.

21. Koyré, *The Astronomical Revolution*, trans. R. E. W. Maddison (Ithaca, NY: Cornell University Press, 1973), p. 42.

22. According to Pascale Delfosse's definition in *Une Idéologie patronale: Essai d'analyse sémiotique* (Paris: Didier, 1974).

23. Science as myth, as form of oppression, etc. Paul Feyerabend goes so far as to deny the possibility of reducing the formation of a hypothesis to a logical process. See Feyerabend, *Against Method* (London: Verso, 1975).

24. Jean-Pierre Vernant, *The Origins of Greek Thought* (Ithaca, NY: Cornell University Press, 1982), pp. 119–29; and *Myth and Thought Among the Greeks* (Boston: Routledge & Kegan Paul, 1983), pp. 223ff.

25. Michel Foucault, *The Order of Things* (New York: Pantheon, 1971); and *Archaeology of Knowledge* (New York: Pantheon, 1972).

26. Gérard Simon, *Kepler astronome astrologue* (Paris: Gallimard, 1979), p. 17.

27. *Ibid.*

28. Paul Ricoeur, *The Rule of Metaphor*, trans. R. Czerny (Toronto: University of Toronto Press, 1975), pp. 244ff.

29. Aristotle, *Poetics*, trans. L. Golden (Englewood Cliffs, NJ: Prentice-Hall, 1968), p. 12 (1450a).

30. Translation proposed by Ricoeur, *Time and Narrative*, trans. K. McLaughlin and D. Pellauer (Chicago: University of Chicago Press, 1984), p. 33. The Greek is *he ton pragmaton sustasis*; Ricoeur's French is *agencement des faits en système.*

31. Aristotle, *Poetics*, p. 16 (1451a).

32. Ricoeur, *The Rule of Metaphor*, p. 39.

33. Umberto Eco, *L'Oeuvre ouverte* (Paris: Éd. du Seuil, 1965), p. 11.

34. Jean Baptiste Crevier, *Rhétorique française* (Paris: Saillant et Dessaint, 1767), vol. 1, p. 32.

35. Stephen Pepper, *World Hypotheses: A Study in Evidence* (Berkeley: University of California Press, 1942). Pepper has applied his classification to art

criticism in *The Basis of Criticism in the Arts* (Cambridge, MA: Harvard University Press, 1963). Hayden White used it for his historiography in *Metahistory: The Historical Imagination in Nineteenth Century Europe* (Baltimore: Johns Hopkins University Press, 1973). Cf. Pepper, "The Root Metaphor Theory of Metaphysics," in W. A. Shibles (ed.), *Essays on Metaphor* (Whitewater, WI: Language Press, 1972), pp. 15–39.

36. Pepper, *Basis of Criticism*, p. 74.

37. *Ibid.*

38. Cf. Ernest Nagel, *The Structure of Science: Problems in the Logic of Scientific Explanation* (London: Routledge & Kegan Paul, 1961), pp. 390–91. Nagel's evaluation of the notions of "wholes," "sum," and "organic unities" is highly critical, even skeptical.

39. Pepper, *Basis of Criticism*, p. 37.

40. *Structure of Science*, p. 411. Nagel is concerned with the "teleological" explanation. The term "formism" does not enter into the discussion.

41. Ernst Mayr, *Evolution and the Diversity of Life* (Cambridge, MA: The Belknap Press of Harvard University Press, 1976), cited by C. Galperin, "Le concept de téléonomie et la notion de programme," in *Modèles et interprétation* (Publication de L'Université de Lille III, 1978), p. 347. We will apply Mayr and Galperin's distinction between teleonomy and teleology to formism. They also discuss a "teleomatic" explanation: the state of equilibrium or highest value of entropy attained non-programmatically in inanimate nature; from our point of view, this is a contextualist schema.

42. Terms used by François Jacob, *La Logique du vivant* (Paris: Éd. du Seuil, 1970), p. 17.

43. Werner Heisenberg, *Physics and Beyond: Encounters and Conversations*, trans. A. J. Pomerans (New York: Harper & Row, 1971), p. 212.

44. Lenoble, *Histoire de l'idée de nature*, p. 245 (see also p. 285). Lenoble articulates the same theme several times in *Mersenne ou la naissance du mécanisme* (Paris: Vrin, 1943).

45. Algirdas Julien Greimas, *On Meaning: Selected Writings in Semiotic Theory*, trans. P. Perron and F. Collins (Minneapolis: University of Minnesota Press,

1987), pp. 20–21.

46. I adopt Eco's distinctions between *system* and *code*, and between *hypo-codification* and *hypercodification*. See Eco, *A Theory of Semiotics* (Bloomington: Indiana University Press, 1978).

47. Giambattista Vico, *The New Science*, trans. of the 3rd ed. (1744) T. G. Bergin and M. H. Fisch (Ithaca, NY: Cornell University Press, 1978), p. 131 (408).

48. *Ibid.* "Since the first men of the gentile world had the simplicity of children, who are truthful by nature, the first fables could not feign anything false; they must therefore have been, as they have been defined above, true narrations."

49. There are numerous pertinent references in Douglas C. Muecke, *The Compass of Irony* (London: Methuen, 1969). Muecke lays too much stress, how-ever, on the idea of the modernity of irony. I will return to this in Chapter One. On irony in the sciences, cf. the remarks of Kenneth Burke, *A Grammar of Motives and A Rhetoric of Motives* (New York: Meridian Books, 1962), pp. 503–17; and White, *Metahistory*, pp. 36–38, *et passim*.

50. Raban Maur, cited by H. Lausberg, *Handbuch der literarischen Rhetorik* (Munich: Hueber, 1960), vol. 1, p. 445.

51. Dante, *Convivio* 2.1 ("Le superne cose de l'etternel gloria").

52. Leibniz, "Tentamen Anagogicum: An Anagogical Essay in the Investi-gation of Causes," in L. E. Loemker (ed. and trans.), *Philosophical Papers and Letters*, 2nd ed. (Dordrecht, Holland: Reidel, 1969), vol. 1, p. 484, n.2; vol. 1, p. 477.

53. Douglas C. Muecke, "Images of Irony," *Poetics Today* 4 (1983), pp. 399ff.

54. Burke, *Grammar of Motives*, pp. 736ff.

55. Jonathan Culler, *The Pursuit of Signs* (London: Routledge & Kegan Paul, 1981), p. 103.

56. This conception of a culture is close to that of J. Lotman and B. Uspenskij, for whom a culture is defined by a double function: it is *memory*, containing the totality of non-hereditary information that is accumulated, kept, and transmitted by a collective entity, and a *mechanism for producing* new state-

ments: *Semiotica e Cultura* (Milan: Ricciardi, 1975), pp. 59-95. I believe less, however, in the global *coherence* of different relations between the transmission and reception of signs that Lotman attributes to a cultural period. As C. Segre remarks, a *reductio ad unum* greatly risks being, in this case, a reduction to zero. See Segre, *Semiotica, Storia et Cultura* (Padua, 1977), p. 19.

57. Michael Riffaterre, "L'intertexte inconnu," *Littérature* 41 (February 1981), p. 6.

58. Severo Sarduy, *Barroco* (Paris: Éd. du Seuil, 1975), p. 7.

59. Heinrich Cornelius Agrippa von Nettesheim, *La Philosophie occulte ou la magie* (Paris: Éd. Traditionnelles, 1962-1963), vol. 1, p. 137 (1.52).

60. Mary B. Hesse, *Models and Analogies in Science* (Notre Dame: University of Notre Dame Press, 1966), p. 171. Cf. P. Ricoeur's discussion, *The Rule of Metaphor*, pp. 239-46.

61. Max Black, "More About Metaphor," *Dialectica* 31 (1977), pp. 441-45.

62. Meyer, *Découverte et justification*.

63. Dedre Gentner, "Are Scientific Analogies Metaphors?" in D. S. Miall (ed.), *Metaphor: Problems and Perspectives* (Atlantic Highlands, NJ: Humanities Press, 1982), pp. 106-32.

64. Michel De Coster, *L'Analogie en sciences humaines* (Paris: Presses Universitaires de France, 1978), p. 6.

65. Hans Blumenberg, *Paradigmen zu einer Metaphorologie* (Bonn: Bouvier, 1960), p. 6.

66. Meyer, *Découverte et justification*, p. 338.

67. Cf. Gérard Genette, "La Rhétorique restreinte," in *Figures III* (Paris: Éd. du Seuil, 1972), pp. 21-40. Genette reacts against similar reductionist tendencies in literary studies.

68. *Irony* will be excluded because it comes into play at another level, as a metatropological figure.

69. Roland Barthes, *The Rustle of Language*, trans. R. Howard (New York: Hill and Wang, 1986), p. 209.

70. Cf. Gérard Genette, *Figures II* (Paris: Éd. du Seuil, 1969), pp. 152-53.

71. See Jean Piaget, *The Construction of Reality in the Child*, trans. M. Cook

(New York: Basic Books, 1954), pp. 350ff.

72. *Ibid.*, p. 352.

73. Meyer (*Découverte et justification*, p. 335) emphasizes the specificity of the return to the characteristic: "Univocality is restored, understood and integrated, if not verified, the metaphor is used up and dissolves in the usage it creates or that predates it."

CHAPTER I: SCIENCE AND IRONY

1. Concerning vertical irony in general, cf. Muecke, "Images of Irony." Omnipotent, omniscient, transcendent, absolute, infinite, and free, God is the opposite of man, conceived as "the archetypal victim...insofar as he may easily be seen trapped and submerged in time and matter, blind, contingent, limited, unfree" (p. 402). Cf. also, by the same author, *The Compass of Irony*, ch. 6, on "general" irony.

2. Plato, *Timaeus*, trans. F. Cornford (New York: Bobbs Merrill, 1959), p. 19 (30a).

3. Plato, *Phaedrus*, trans. W. Hamilton, in *Phaedrus and the Seventh and Eighth Letters* (New York: Penguin, 1973), p. 97 (275).

4. Ptolemy, *Almagest*, trans. R. Catesby Taliaferro, in *Great Books of the Western World*, R. M. Hutchins (ed.) (Chicago: University of Chicago Press, 1952), vol. 16, ch. 13.2. If Ptolemy accepted Aristotle's physics, his Platonism was nonetheless profoundly real. Cf. especially Pierre Duhem, *Le Système du monde* (Paris: Hermann, 1954), vol. 1, p. 496; Jürgen Mittelstrass, *Die Rettung der Phänomene* (Berlin: De Gruyter, 1962), pp. 168–69; and I. Dambska, "L'Epistemologie de Ptolemée," in *Avant, avec, après Copernic*, Centre International de Synthèse (Paris: Blanchard, 1975), pp. 31–37.

5. Cf. Muecke ("Images of Irony," especially p. 403), who distinguishes, in addition to the fundamental opposition (high-low), a series of associated oppositions (light-dark, mobile-immobile, one-many, reality-illusion, meaning-absurdity, happiness-misery...).

6. Northrop Frye, *Anatomy of Criticism* (Princeton: Princeton University Press, 1957), p. 34.

7. Ptolemy, *Almagest* 1, Preface: "And indeed this same discipline would more than any other prepare understanding persons with respect to nobleness of actions and character...making its followers lovers of that divine beauty, and making habitual in them, and as it were natural, a like condition of the soul." Cf. Plato's praise of astronomy in *Laws* 7.

8. *Ibid.* 13.2.

9. *Ibid.* 3.4.

10. *Ibid.* 13.2.

11. I borrow this expression from Dambska, "L'Epistemologie de Ptolemée," p. 34.

12. Cited in Pierre Duhem, *To Save the Phenomena*, trans. E. Doland and C. Maschler (Chicago: University of Chicago Press, 1969), p. 11. Geminus' text also appears in Simplicius, *In Aristotelis physicorum libros quattuor priores commentaria*. In *To Save the Phenomena*, originally published in 1908 ("Sozein ta phainomena," *Annales de philosophie chrétienne* 6, pp. 113–39, 277–302, 352–77, 482–514, 561–92), Duhem described the development of an instrumentalist conception of mathematical astronomy. The historical aspect of the subject has been taken up by Mittelstrass in *Die Rettung der Phänomene*. The publication in 1969 of Doland and Maschler's English translation of Duhem has stimulated numerous critical discussions. See especially John L. Heilbronn, "Duhem and Donahue," in R. S. Westman (ed.), *The Copernican Achievement* (Berkeley: University of California Press, 1975), pp. 276–84; Geoffrey E. R. Lloyd, "Saving the Appearances," *Classical Quarterly* 28 (1978), pp. 202–22; Nicholas Jardine, "The Forging of Modern Realism: Clavius and Kepler Against the Sceptics," *Studies in the History and Philosophy of Science* 10 (1979), pp. 141–73; and also by Jardine, *The Birth of History and Philosophy of Science* (Cambridge: Cambridge University Press, 1984).

13. Kenneth Burke, *Grammar of Motives*, p. 512: "Irony arises when one tries, by the interaction of terms upon one another, to produce a development which uses all the terms."

14. Cf. Michel Lerner, "L'Humanisme a-t-il sécreté des difficultés au développement de la science au XVIᵉ siècle? Le cas de l'astronomie," *Revue de*

*synthèse* (1979), p. 55: "The divorce between physics and astronomy was thus consummated with Ptolemy, and medieval authors up to and including Regiomontanus had already accommodated the division that Geminus formulated theoretically in the first century B.C."

15. Cf. Edward Grant, "Cosmology," in D. C. Lindberg (ed.), *Science in the Middle Ages* (Chicago: University of Chicago Press, 1978), p. 280. Concerning the translation of scientific works in general, cf. Lindberg's essay in this work, pp. 52–90.

16. Thomas Aquinas' commentary on *De caelo* (*In libros Aristotelis de caelo et mundo exposito* 2.17): "As for these astronomers, the hypotheses that they have invented are not necessarily true: in fact, even though they appear to establish the truth of these hypotheses, we must not say that they are necessarily true, for it may be that phenomena of the stellar world can be saved in some other manner that men have not yet understood." Cf. also Aquinas, *Summa theologica* 1.32.1–2. For Averroës, cf. *Aristotelis metaphysicorum libri XIV cum Averrois...in ea opera omnia commentarii* (Venice, 1550–1552), vol. 8, fol. 158ᵛ. The passage from *De caelo* (297a2–6) is cited by Rheticus in the *Narratio prima* (3CT, p. 194): "We have evidence for our view in what the mathematicians say about astronomy. For the phenomena observed as changes take place in the figures by which the arrangement of the stars is marked out occur as they would on the assumption that the earth is situated at the center."

17. Hans Blumenberg, *Genesis of the Copernican World* (Cambridge, MA: MIT Press, 1987), p. 220.

18. Nicole Oresme, *Le Livre du ciel et du monde*, A. D. Menuet and A. J. Denomy (eds.), trans. A. D. Menut (Madison: University of Wisconsin Press, 1968), p. 354.

19. Oresme, *Le Livre*, pp. 537–39.

20. Duhem, *Le Système du monde*, vol. 9, p. 341.

21. In the Introduction to Oresme, *Le Livre*, p. 27.

22. "Renaissance Science as Seen by Burckhardt and His Successors," in T. Helton (ed.), *The Renaissance* (Madison: University of Wisconsin Press, 1961), p. 92.

23. Oresme, *Le Livre*, p. 531. The passage proposes an extension of the accepted principle of *accommodatio*. Concerning the *Genesis* account, cf. Aquinas, *Summa theologica* 1.68.3.

24. On the notion of "evidence," cf. Anneliese Maier, *Ausgehendes Mittelalter* (Rome: Ed. di Storia e Letteratura, 1967), vol. 2, pp. 367–418.

25. Cf. Heiko Oberman, "Reformation and Revolution," in O. Gingerich (ed.), *The Nature of Scientific Discovery* (Washington, D.C.: Smithsonian Institution Press, 1975), pp. 134–69. Amos Funkenstein notes that the distinction between *potentia absoluta* and *potentia ordinata*, already present in Aquinas, grew deeper with the nominalists. See Funkenstein, "The Dialectical Preparation for Scientific Revolutions," in Westman, *The Copernican Achievement*, pp. 182–83.

26. Cf. Duhem, *To Save the Phenomena*; Maier, *Studien zur Naturphilosophie der Spätscholastik* (Rome: Ed. di Storia e Letteratura, 1949); Alistair C. Crombie, *Robert Grosseteste and the Origins of Experimental Science (1100–1700)* (Oxford: Clarendon Press, 1957); and John H. Randall, Jr., *The School of Padua and the Emergence of Modern Science* (Padua: Antenore, 1961). These are only a few indicative references among a vast literature.

27. Cf. especially Koyré, *Études d'histoire de la pensée scientifique*, pp. 61–86. See Paul L. Rose, *The Italian Renaissance of Mathematics* (Geneva: Droz, 1975), p. 4; and Lerner, "Le Cas de l'astronomie," p. 16, n.1, for selective bibliographies on the issue.

28. Oresme, *Le Livre*, p. 365. As Funkenstein ("The Dialectical Preparation for Scientific Revolutions") notes, the insistence on divine liberty effectively transformed the problems raised by Aristotle concerning the mode of the possible or the impossible into questions of probability or improbability.

29. *Impetus* "is always represented as some kind of power or force which passes from the mover to the *moved*, and maintains motion or, better yet, is the cause of motion" (Koyré, *Études*, pp. 180–81). Since there is no resistance in the heavens, motion there can continue indefinitely. On the significance of the theory of impetus, cf. Maier, *Ausgehendes Mittelalter*, vol. 1, pp. 353–80.

30. Jean Buridan, *Quaestiones super libris quattuor de caelo et mundo*, E. A. Moddy (ed.) (Cambridge, MA: The Medieval Academy of America, 1942),

pp. 180–81 (2.12). Buridan rejects the notion that the earth moves.

31. Cf. Anneliese Maier, *Die Vorlafer Galileis im 14. Jahrhundert* (Rome: Ed. di Storia e Letteratura, 1949), p. 101, n.41.

32. Cf. Maier, *Ausgehendes Mittelalter*, vol. 2, pp. 398ff.

33. *Ibid.*, vol. 2, p. 403: "Natural science can – and must – dismiss the possibility of miracles; accordingly it can confine itself to the *influentia generalis* with which God, *de potentia ordinata*, directs natural events, and disregard the possibilities that flow from his *potentia absoluta*. In short, this leads to the postulate of a methodological split between theology and natural science."

34. Frye, *Anatomy of Criticism*, p. 214.

35. Erwin Panofsky has emphasized the complementary relationship between nominalism and mysticism in *Gothic Architecture and Scholasticism* (New York: New American Library, 1957), pp. 14ff.: "In fact, these two extremes, mysticism and nominalism, are, in a sense, nothing but opposite aspects of the same thing."

36. Cf. Georg Lukács, *Theory of the Novel* (Cambridge, MA: MIT Press, 1971), p. 90. It is true that the author reserves negative mysticism to "times without a god." But the expression can designate the inexistence just as well as the removal of God. It is evidently the second meaning that applies here. Concerning the idea of "homogeneous" epochs, on the other hand, it is appropriate to substitute the idea of discursive practice or diversified discursive domains within a given epoch. This applies to the complementarity between nominalist discourse and mystical discourse, or between astronomical discourse – considered as a practice carried out in human solitude, far from God – and cosmological discourse (cf. Dante) founded on faith or *auctoritas*, self-assured and favorable to mysticism. On the importance of irony in the Middle Ages, cf. Edmund Reiss, "Medieval Irony," *Journal of the History of Ideas* 42 (1981), pp. 209–26.

37. On skepticism during the Renaissance, cf. Richard H. Popkin, *The History of Scepticism from Erasmus to Descartes* (Assen: Van Gorcum, 1960); and Charles B. Schmitt, "The Recovery and Assimilation of Ancient Scepticism in the Renaissance," *Rivista critica di storia della filosofia* 27 (1972), pp. 363–84.

38. Cf. Oberman, "Reformation and Revolution."

39. Cf. Jardine, "The Forging of Modern Realism." Pico della Mirandola had already drawn Plato and Aristotle into the sphere of skepticism in his *Examen vanitatis doctrinae gentium et veritatis christianae doctrinae*.

40. Georg von Peurbach, *Theoricae novae planetarum* (Cologne: Birckmans, 1591), unpaginated preface: "In order to satisfy the experience [of their senses], the Astronomers are obliged to assign so monstrous a form to the celestial spheres, that if they speak truly, it would be impossible to think of anything more deformed than Heaven.... And nonetheless, reason refuses to admit that such a deformity could be attributed to the very noble celestial bodies.... One can therefore learn from Astronomy that the blindness of human intelligence is so great that it is not even capable of reconciling diverse sense experiences."

41. Michel de Montaigne, "Apology for Raymond Sebond, in *The Complete Works*, trans. D. M. Frame (Stanford: Stanford University Press, 1957), pp. 371–72. Montaigne follows Sextus Empiricus' *Outlines of Pyrrhonism* (1.1.2–3).

42. If I know that I know nothing, it is not true that I know nothing.... Montaigne (p. 392) emphasizes the similarity to the paradox of the liar. Cf. also Alexandre Koyré, *Épiménide le menteur* (Paris: Hermann, 1947), p. 36.

43. Thus Pyrrho "wanted to make himself a living, thinking, reasoning man..." (Montaigne, p. 374). Cf. Jean Paul Dumont, *Le Scepticisme et le phénomène* (Paris: Vrin, 1972, pp. 43, 237–38, and A. Tournon, *Montaigne: La Glose et l'essai* (Lyons: Presses Universitaires de Lyon, 1983), p. 245.

44. For Kepler's testimony, cf. his *Apologia contra Ursum* and preface to *Astronomia nova*, where he replies to Ramus, who had severely criticized the *Ad lectorem* in a letter to Rheticus, translated by Marie Delcourt in "Une lettre de Ramus à Joachim Rheticus (1563)," *Bulletin de l'Association Guillaume Budé* 41 (1934), pp. 3–15; cf. also E. Rosen, "The Ramus-Rheticus Correspondence," *Journal of the History of Ideas* 1 (1940), pp. 363–68; and the copiously annotated translation in Rheticus *Narratio prima*, H. Hugonnard-Roche and J.-P. Verdet (eds. and French trans.), with the collaboration of M.-P. Lerner and A. Segonds (Wroclaw: Maison d'édition de l'Academie polonaise des sciences, 1982), pp. 238–45. The text by Praetorius (letter to Herwart von Hohenburg) was

published by Ernst Zinner, *Entstehung und Ausbreitung der copernicanischen Lehre* (Erlangen: Mencke, 1943), p. 454.

45. Cited in Hugonnard-Roche and Verdet (eds.), *Narratio prima*, pp. 216–17.

46. Arthur Koestler, *The Sleepwalkers* (Harmondsworth: Penguin Books, 1964), pp. 169–78.

47. Oberman, "Reformation and Revolution"; and B. Wrightsman, "Andreas Osiander's Contribution to the Copernican Achievement," in Westman, *The Copernican Achievement*, pp. 213–43. It is doubtful that Osiander intended his *Ad lectorem* to appear as an autograph text. On several occasions, the author of *De revolutionibus* is designated in the third person: "authorem huius operis," "hic artifex." Moreover, Copernicus' own letter-preface (addressed to the Pope) is clearly distinguished, in the first edition, by the usual title *Praefatio authoris*, which is lacking above Osiander's *Ad lectorem*.

48. Cf. especially OR, p. 3: Copernicus asserts that his endeavor as a "philosopher" is "to seek the truth in all things"; "...if I made the opposite assertion that the earth moves"; OR, pp. 42–43: Copernicus protests "the confusion in the astronomical traditions concerning the derivation of the motions of the universe's spheres," the absence of "certainty" in the explanation of "the movements of the world machine"; OR, p. 45: Copernicus confidently asserts, "I have no doubt that acute and learned astronomers will agree with me if, as this discipline especially requires, they are willing to examine and study, not superficially but thoroughly, what I adduce in this volume in proof of these matters."

49. Cited in French by Tournon, *Montaigne*, p. 210.

50. *Ibid.*

51. *Ibid.* Tournon furnishes other examples of the same kind. On the *declamatio* in the Renaissance in general, see M. Van der Poel, *De declamatio bij in humanisten* (Niewkoop: Der Graaf. 1987).

52. Praetorius insists on the contradiction between the two texts and declares that Copernicus' own opinion comes out clearly in the dedicatory epistle addressed to the Pope. See Zinner, *Entstehung*, p. 454.

Chapter II: Science and Anagogy

1. Fracastoro, *Homocentrica* (Venice, 1538), fol. 1ʳ⁻ᵛ.

2. As noted in the Zeller edition of *De revolutionibus*, this passage closely follows the text of Pliny's *Natural History* 2.3. See F. and K. Zeller (eds.). *Nicolaus Kopernikus Gesamtausgabe* (Munich: Oldenbourg, 1949), vol. 2, p. 345

3. Cf. the textual comparisons in the Klaus-Birkenmajer edition of Book 1 of *De revolutionibus* (Berlin: Akademieverlag, 1959), p. 101, nn. 25–27.

4. Marsilio Ficino, *Théologie platonicienne*, R. Marcel (ed. and trans.) (Paris: Les Belles Lettres, 1964), vol. 2, p. 226 (13.14). In *De vita triplici* 3.19, Ficino expresses his admiration for an astronomical clock constructed by Lorenzo della Volpaia, which reminds him of Archimedes' clock. Poliziano describes this clock in detail in a letter to Francesco della Casa. See Poliziano, *Opera* (Venice: Manuce, 1498), fol. 1ʳ⁻ᵛ.

5. Cf. Hans Blumenberg, *Paradigmen zu einer Metaphorologie* (Bonn: Bouvier, 1960), ch. 6; and Jean Dagens, "Hermétisme et cabale en France de Lefèvre d'Etaples à Bossuet," *Revue de littérature comparée* 35 (1961), pp. 5-16.

6. Cf. Thomas S. Kuhn, *The Copernican Revolution* (Cambridge, MA: Harvard MIT Press, 1957), p. 202.

7. Cf. Hans Blumenberg, *Genesis of the Copernican World* (Cambridge, MA: MIT Press, 1987), p. 202.

8. Lorenzo Valla, *Opera omnia* (Basel: Fröben, 1540), vol. 1, p. 659. And also: "we see what great benevolence the creator of the universe has demonstrated to us, who created in its entirety the very universe *for man*...in truth, we would be altogether lacking in reason if we did not recognize that this work has been made *for men* by God. So that you understand how much greater your power is, you for whom, as you see, the entire universe, and all the heavens were created, *for you alone*, I say" (*ibid.*, vol. 1, p. 982 [my italics]).

9. Charles Bovelles, *De Sapiente*, ed. R. Klibansky, in Ernst Cassirer, *Individuum und Kosmos in der Philosophie der Renaissance* (Leipzig: Teubner, 1927), pp. 353-54: "In fact, if you place all of creation around the circumference of the universe (like around the firmament), you will have to place Man in the middle and at the center; from there, the entire circumference of the universe

will appear brightly transparent and will be revealed to him. But if you place all of creation along the base of a triangle, you will have to place Man at the apex of the triangle, toward which the base flows along the two sides, toward which the entire surface of the triangle flows, and from which the entire base can be seen and easily observed all at once." On the theme of the microcosm in Bovelles, cf. my study: "Le Microcosme ou l'incomplétude de la représentation," *Romanica Gandensia* 17 (1980), pp. 183–92.

10. Cf. Blumenberg's commentary, *Genesis*, pp. 206–08. The copies of Bovelles' works that Copernicus read and annotated are kept at Uppsala: *Liber de intellectu, Liber de sensu*...(Paris, 1511).

11. Cited in Rheticus, *Narratio prima*.

12. Tycho Brahe calls Copernicus a "new Ptolemy." See Brahe, *Opera omnia*, ed. J. L. E. Dreyer (Copenhagen: 1913–1929), vol. 1, p. 149. Clavius presents him as "extraordinary restorer of Astronomy in this our age" [*nostro hoc saeculo Astronomiae restitutor egregius*]. See Clavius, *In sphaeram...commentarius* (Lyons: Gabian, 1593), p. 67. Likewise, Reinhold calls Copernicus a second Atlas and a second Ptolemy. See Edward Rosen, "Studia Copernicana," *Journal of the History of Ideas* 35 (1974), p. 524. Bruno employs the metaphor of a new light. See Bruno, *De immenso*, in *Opera latine conscripta*, ed. F. Fiorentino (Naples: Morano, 1889), vol. 1, p. 1. Book 3, ch. 9 is entitled "De lumine Nicolai Copernici," and contains this passage (p. 381): "It is wonderful, O Copernicus, that from the deep blindness of our age, where the light of philosophy and of the sciences that flow from it was extinguished, you could emerge...."

13. Cf. also Rheticus, *Narratio prima* (3CT, p. 186): "Furthermore, concerning my learned teacher I should like you to hold the opinion and be fully convinced that for him there is nothing better or more important than walking in the footsteps of Ptolemy and following, as Ptolemy did, the ancients and those who were much earlier than himself." On Copernicus' fidelity to the numerical data of the Ancients, cf. Otto Neugebauer, "On the Planetary Theory of Copernicus," *Vistas in Astronomy* 10 (1968), pp. 89–113.

14. The manuscript of *De revolutionibus* contains a reference to Aristarchus that was dropped in the printed text. On the ancient "precursors" of Coperni-

cus, cf. the bibliographical indications furnished by Koyré, *The Astronomical Revolution*, p. 71, n.3; and the more recent contribution of O. Jakob, "Die Reception des 'antiken Copernicus' Aristarch von Samos in Antike und Neuzeit," *Anregung* 29 (1983), pp. 299–314.

15. Ptolemy, *Almagest* 1.1 *in fine.*

16. 3CT, p. 108. The phrase is taken up again at the end of the text (3CT, p. 187). Eugenio Garin notes that the *Didaskalikos* had been translated by Pietro Balbi (for Nicholas of Cusa) and by Ficino. See Garin, *Rinascite e rivoluzioni* (Bari: Laterza, 1975), p. 278.

17. Cf. Gérard Genette, *Palimpsestes* (Paris: Éd. du Seuil, 1982), p. 14: "I call hypertext any text derived from a previous text by simple ... or ... indirect transformation." Under the headings of *hypertextuality* and *paratextuality* (cf. below, n.19), Genette points to varieties of *intertextual* (for Genette *transtextual*) relations in general.

18. It is too bad that Genette pays relatively little attention to the Renaissance. As an exemplary marriage of *renovatio* and *innovatio*, hypertextuality corresponds to a major textual function during that period. Thomas M. Greene has studied in depth the relation of "texts" to their "subtexts." See Greene, *The Light in Troy* (New Haven: Yale University Press, 1982).

19. On the notion of "paratext," cf. Genette, *Palimpsestes*, pp. 9–10; and *Seuils* (Paris: Éd. du Seuil, 1987).

20. Cf. Edward Grant, "Cosmology," in D. C. Lindberg (ed.), *Science in the Middle Ages* (Chicago: University of Chicago Press, 1978), p. 267: "it is important to recognize that the virtues of such an approach, which emphasized thorough and systematic analysis of distinct problems in a prescribed order, were offset by the absence of any cohesive integration of the many separately derived conclusions as well as a failure to detect inconsistencies between questions treated within the same treatise."

21. Derek J. Price, "Contra-Copernicus," in M. Clagett (ed.), *Critical Problems in the History of Science* (Madison: University of Wisconsin Press, 1969), p. 215. The order adopted by Ptolemy already corresponded to a canonical model. Cf. Pierre Duhem, *Le Système du monde* (Paris: Hermann, 1954), vol. 1,

pp. 468-77. Cf. also Marie Boas, *The Scientific Renaissance* (New York: 1962), p. 74. On Copernicus' deviations from Ptolemy's formal model, cf. Jerome R. Ravetz, *Astronomy and Cosmology in the Achievement of Nicolaus Copernicus* (Warsaw: Polish Academy, 1965), p. 55.

22. Cf. Greene, *The Light in Troy*, pp. 38-39.

23. Cf. Genette, *Palimpsestes*, p. 372. The term "transmotivation" is appropriate for Copernican hypertextuality; Copernicus adopted an existing form of exposition but subjected it to a different motivation: his construction was heliocentric rather than geocentric.

24. 3CT, p. 147: "By what number could anyone more easily have persuaded mankind that the whole universe was divided into spheres by God that Author and Creator of the world?" A perfect number is equal to the sum of its divisors other than itself ($6 = 1 + 2 + 3$). In *Mysterium cosmographicum*, Kepler would seek geometric justification for the number of planets. When Huyghens discovered the first moon of Saturn, he thought he had discovered the last link in the creation: the moon, Jupiter's four satellites and Saturn's one satellite also add up to six, such that the cosmos comprises 6 planets and 6 satellites, 12 in all – the symbolic number of completeness. Cf. I. Bernard Cohen, "Perfect Numbers in the Copernican System: Rheticus and Huyghens," in *Science and History: Studies in Honor of E. Rosen* (Wroclaw: The Polish Academy of Sciences Press), pp. 419-25; and by the same author, *The Newtonian Revolution* (Cambridge: Cambridge University Press, 1980), pp. 20-21.

25. Cf. OR, pp. 25-26. We do not know who was responsible for suppressing the letter from the printed text. In any case, the fact that Copernicus had first chosen to close Book 1 by such a long Pythagorean text shows his interest in Pythagoreanism. The question of the precise importance of Pythagoreanism in Copernicus' work is controversial. Cf. especially: Thomas W. Africa, "Copernicus' Relation to Aristarchus and Pythagoras," *Isis* 52 (1961), pp. 403-09; Edward Rosen, "Was Copernicus a Pythagorean?" *Isis* 53 (1962), pp. 504-08 (and Africa's reply, p. 509); William Stahlman, "On Recent Copernicana," *Journal of the History of Ideas* 34 (1973), pp. 483-89; Stanislaw Mossakowski, "The Symbolic Meaning of Copernicus' Seal," *ibid.*, pp. 451-60; and B. Bilinski, *Il*

*Pitagorismo di Nicolo Copernico* (Wroclaw: The Polish Academy of Sciences Press), 1977. In short, whereas Copernicus' reticence seems similar to and even modeled on the reticence of the Pythagoreans, the influence of their cosmology on the *De revolutionibus* is uncertain. Of course the first point explains the second.

26. Josse Clichtove, "Preface," in Jacques Lefèvre d'Étaples, *In quoscunque philosophiae naturalis libros Paraphrasis* (Paris: Regnault, 1525): "But in fact these chimerical and feeble arguments, which run counter to scientific truths, ordinarily do not lead to an understanding of the certitude and integrity of those truths, but rather turn away from them in favor of petty sophistical distinctions, bearing no relation to true doctrine."

27. Latomus, *Pro dialogo de tribus linguis apologia*, cited by André Godin, "Fonction d'Origène dans la pratique exégétique d'Erasme...," in Olivier Fatio and Pierce Fraenkel (eds.), *Histoire de l'exégèse au XVIᵉ siècle* (Geneva: Droz, 1978), p. 37.

28. Cf., for example, the respect expressed by Lefèvre d'Étaples, *In quoscunque, Prologus in physicos libros Aristotelis*, folio cclv: "In effect, I testify to such interest in all the Peripatetics and especially to the great Aristotle, chief of all the true philosophers, that if I draw some knowledge from their disciplines that I judge useful, beautiful, and holy, I want it to be communicated to everyone, so that everyone is overcome with admiration for the passion of these philosophers and shows them the veneration they merit, and loves them."

29. John G.A. Pocock, *Politics, Language and Time* (New York: Atheneum, 1971), pp. 255–56. The opposition to the humanists, especially within the Church, corresponds to the opposition Pocock describes of "traditionalists" to "classicists" (p. 254): "The traditionalist...will always distrust the classicist, seeing in him the well-meaning author of a potentially radical doctrine."

30. Mythical relationship: the cyclical conception of history or regeneration following degradation; images of initiation or "sleep" or "death" in the "shadows" of the Middle Ages followed by an "awakening" or "birth" to a new and ancient "light"; the cult of "strong" time, whose relics are venerated; ritual repetition of the ancient forms in art and literature; the educational func-

tion attributed to the lives recounted by Plutarch or Livy, etc. On all these points, the Renaissance recalls the "lived" experience of myths, as described by Mircea Eliade, especially in *Myth and Reality*, trans. W. R. Trask (New York: Harper & Row, 1963), which contains allusions to the Renaissance, especially pp. 135–36 and 156–57.

31. Cf. Eugenio Garin, *L'Età nuova* (Naples: Morano, 1969), pp. 493ff.; and Rose, *The Italian Renaissance of Mathematics*, who describes in detail the relations between humanists and mathematicians in sixteenth-century Italy. We hardly need emphasize, in this connection, the importance of the printing press. Cf. Owen Gingerich, "Copernicus and the Impact of Printing," in Arthur Beer and K. Aa. Strand (eds.), *Copernicus: Yesterday and Today* (New York: Pergamon Press,1975), pp. 201–07; and the well-known general references (Febvre and Martin...); and especially the book by Elizabeth L. Eisenstein, *The Printing Press as an Agent of Change* (Cambridge: Cambridge University Press, 1979), vol. 2, ch. 7 ("Resetting the Stage for the Copernican Revolution"), pp. 575–635.

32. Poliziano, *Miscellaneorum centuria secunda*, ed. V. Branca and M. P. Stocchi (Florence: Olschki, 1978), p. 3, Branca writes of the "almost sacred dignity of the true science of man" which this passage confers on philology. See Branca, *Poliziano e l'umanesimo della parola* Turin: Einaudi, 1983), p. 157.

33. *The Epistles of Erasmus*, trans. F. M. Nichols, vol. 1 (London: Longmans, Green and Co., 1901), p. 382 (Epistle 182). The passage is from a letter to Christopher Fischer, which served as a preface to Valla's *Adnotationes* on the New Testament. On the Polish contacts and influence of Erasmus, especially concerning Copernicus' circle, cf. Andrzej Kempfi, "Érasme et la vie intellectuelle en Pologne au temps de Nicolas Copernic," in *Colloquia Erasmiana Turonensia* (Paris: Vrin, 1972), vol. 1, pp. 397–406; and Reijer Hooykaas, *Rheticus' Treatise on Holy Scripture and the Motion of the Earth* (Amsterdam: North Holland, 1984), ch. 2, pp. 20ff. Interesting data also appears in Henri de Vocht, *Danticus and his Netherlandish Friends* (Lovain: Vandermeulen, 1961).

34. Erasmus' deference and humility may be feigned. Theology does not ask grammar to serve her. Grammar seeks to impose a relation that obligates

the "mistress" to acknowledge her existence and importance. Guillaume Budé did not hesitate to call philology a *scientia orbicularis* (all-embracing science). See Budé, *De philologia. De studio litterarum*, preface by August Buck (Stuttgart: Fromman, 1964), p. 19.

35. Erasmus, *Opera omnia*, J. Clericus (ed.) (Leyde: Vander Aa, 1703-1706), vol. 6, cols. 926-28.

36. J. Plattard, "L'Écriture Sainte dans l'oeuvre de Rabelais," *Revue des études rabelaisiennes* 8 (1911), pp. 298-99. Plattard includes other examples of the usage of the term *mataiologoi*.

37. Erasmus, *Opera omnia*, vol. 2, col. 355.

38. The use of the term "paradigm" does not necessarily imply (nor does it exclude) a relationship of influence: perspective constitutes an *example*, a *paradeigma* that particularly illuminates a sign's new relation — whether it is an example that one has followed or simply one among several manifestations, without the status of "cause." In an attempt to describe and not explain, I do not choose between the two possibilities.

39. Timothy J. Reiss, *The Discourse of Modernism* (Ithaca, NY: Cornell University Press, 1982), p. 31.

40. Cf., however, *ibid.*, pp. 37, 329, and *passim* (brief allusions to perspective). Reiss also edited volume 49 of *Yale French Studies*, entitled *Science, Language and the Perspective Mind* (New Haven: Yale University Press, 1973).

41. Émile Benveniste, *Problèmes de linguistique générale*, vol. 2 (Paris: Gallimard, 1980), p. 223.

42. *Ibid.*, pp. 224ff. Cf. also ch. 3, pp. 43ff.

43. Leon Battista Alberti, *De pictura, de statua*, C. Grayson (ed. and trans.) (London: Phaidon, 1972), p. 56.

44. *Ibid.*

45. It is well known that the perspectivist technique effectively corresponds to a reductive conception of real seeing: it presupposes monoscopic, fixed vision in an entirely homogeneous space. Cf. especially Erwin Panofsky, *La Perspective comme forme symbolique* (Paris: Éd. de Minuit, 1975), pp. 42ff. (The English translation, *Perspective as Symbolic Form*, is forthcoming from Zone Books.)

46. In Plato's terms, perspective presents "simulacra," not copies (*Sophist* 235d–236c).

47. Alberti, *De pictura*, p. 54: "First I inscribe on the painting surface a rectangle as large as I want, which will act for me as an open window through which events can be observed...." It is worth noting, in these lines, the insistence on the subject's sovereignty ("as large as *I* want, which will act *for me*").

48. Cited by John White, *The Birth and Rebirth of Pictorial Space* (London: Faber & Faber, 1957), p. 127. White comments, "It is the particular virtue of the new perspective that it contains within itself the true principles of nature. The processes of measurement and comparison, which alone give meaning to what is seen, attain a mathematical precision beyond the reach of the fallible human eye and mind."

49. In this manner, artistic perspective becomes truly exemplary or paradigmatic. Cf. Panofsky, *La Perspective*, pp. 157ff.

50. Cf. Panofsky, *Renaissance and Renascences in Western Art* (New York: Harper & Row, 1972), p. 108: "In the Italian Renaissance the classical past began to be looked upon from a fixed distance, quite comparable to the 'distance between the eye and the object' in...focused perspective." Cf. also p. 111: in the Middle Ages, "for want of a 'perspective distance' classical civilization could not be viewed as a coherent cultural system within which all things belonged together."

51. Valla, *Dialecticae disputationes*, in *Opera omnia* (Basel, 1540/Turin, 1962), vol. 1, p. 649: "Assuredly, when we affirm that something is true or false, this affirmation relates to the mind of the person who is speaking.... That is why it is in us, that is, in our mind, that truth or falsity is found."

52. Here is the complete sentence: "This is why there is no difference when we say what wood is, what stone is, iron, or man, and when we say what 'wood,' 'stone,' 'iron,' or 'man' means...." (*Opera omnia*, vol. 2, p. 677). On Valla's theory of language, cf. Hanna Barbara Gerl, *Rhetorik als Philosophie: Lorenzo Valla* (Munich: Fin, 1974); and Richard Waswo, "The 'Ordinary Language Philosophy' of Lorenzo Valla," *Bibliothèque d'humanisme et Renaissance* 41 (1979), pp. 255–71. Both studies emphasize the importance of the passages cited here.

53. Cf. the introduction to Book 2 of *De voluptate* (*Opera omnia*, vol. 1, pp. 927ff.): "But let us recall that this [abundance of speech] is the foundation of most things."

54. Cf. Gerl, *Rhetorik*, p. 65: "For Valla, on the other hand, speech comes into play only in relation to things, and especially the mastery of things necessary for humans. Thus, speech is not a representation or imitation of existing things, rather it brings them into existence in the first place. For Valla, speech is the second, specifically human creation of the world, the design for reality. Reality is thus no longer the mere thing (as in nominalism), nor the mere abstraction (as in realism), but rather the thing as explained by the word, and fixed in one meaning for man, out of the chaotic multiplicity of possible meanings."

55. Cf. Waswo, "The Reaction of Juan Luis Vivès to Valla's Philosophy of Language," *Bibliothèque d'humanisme et Renaissance* 42 (1980), pp. 595–609. The citations that follow are taken from this study.

56. Vivès, *De disciplinis libri XX* (Anvers, 1531), vol. 3, p. 193.

57. *Ibid.*, vol. 2, p. 194.

58. *Ibid.*

59. *Ibid.*, vol. 3, pp. 378–79.

60. Immanuel Kant, *Critique of Pure Reason*, trans. N. K. Smith (London: Macmillan, 1934), p. 19n (from "Preface to the Second Edition").

CHAPTER III: "THE PRINCIPAL CONSIDERATION..."

1. "Intervention," in *Avant, avec, après Copernic*, p. 136.

2. According to the typology established by Rudolf Allers, "Theories of Macrocosms and Microcosms," *Traditio* 2 (1944), pp. 319–407. Copernicus' comparison belongs, on the other hand, in the category that Allers calls "structural microcosmism." To the degree, moreover, that it presents the universe as divine work and pictorial representation as human work, it also recalls the "aesthetic" variant of Allers' "holistic" microcosmism.

3. Cf., for example, Heinrich Cornelius Agrippa von Nettesheim, *La Philosophie occulte ou la Magie*, vol. 3, p. 155 (1.52): "The elements are in man according to the true properties of their nature. There is in man a kind of ethereal

body, vehicle of the soul, that represents the heavens. There is in man the vegetative life of plants, the animal senses...."

4. Aristotle, *De caelo*, in *Works of Aristotle*, trans. J. L. Stocks (Oxford:
Clarendon Press, 1930), vol. 2, 2.10.291a.

5. *De revolutionibus* 1.10 (OR, pp. 18–22). For a table synthesizing the pro-
posals, see E. Maula, *Studies in Eodoxus: Homocentric Spheres* (Helsinki: Societas
Scientiarum Fennica, 1974), p. 23.

6. Cf. Koyré, *The Astronomical Revolution*, pp. 51–52: "Although Ptolemy
does indeed put forward determinations of distance, he gives them, so to speak,
only by way of extra information. In fact, Ptolemy's technical astronomy is based
on the calculation of angles (and not of distances); consequently, the order of
the celestial bodies could be reversed, and they could be placed anywhere or
at any distance in the firmament (always provided that the relationship between
the length of the radii of the deferents and the epicycles be maintained)."

7. See, for example, Pontus de Tyard's long and elegant evocation of the
celestial dance in J. C. Lapp (ed.), *L'Univers* (Ithaca, NY: Cornell University
Press, 1950), pp. 24–27.

8. In the sixteenth century, Amico and Fracastoro attempted to revive
homocentric astronomy. See Amico, *De motibus corporum coelestiu* (1536); and
Fracostoro, *Homocentrica* (Venice, 1538).

9. For a presentation of the problem, see Kuhn, *The Copernican Revolution*,
ch. 2. Concerning the equant, see below Chapter 4.

10. Otto Neugebauer calls attention to the fact that Pierre Duhem's description of the moon's motion in *Le Système du monde* vol. 1, pp. 494ff. is incorrect, even though it has often been cited. See Neugebauer, *The Exact Sciences
in Antiquity* (New York: Harper, 1962), p. 206.

11. *Homocentrica* 2.26, fol. 64$^v$.

12. Cf. Pontus de Tyard (*L'Univers*, p. 105): "Beyond this, the proportion
of the epicycles seemed very improper to him [Copernicus], even the epicycle
of Venus, imagined by the earliest astronomers as so large in diameter that the
body of this planet, when it is at the perigee, & lowest in the heavens, often
appeared sixteen times larger than when it was higher at its apogee; it even

descended below the the crescent of the Moon into the region of the elements. This accident is so contrary to nature, & to truth...." We should emphasize that for Copernicus Venus' epicycle remained very large. Osiander's statement on this point (OR, p. xvi) is ambiguous. Does he point to a defect in the Ptolemaic representation, or does he imply that the defect persists in Copernicus?

13. This fact was mentioned by Peurbach and Regiomontanus in *Epitoma* 5.22, which Copernicus obtained as early as 1496 or 1497. See A. Birkenmaier, in *Studia Copernicana* 5, pp. 602–604.

14. See Kuhn, *The Copernican Revolution*, pp. 64–72.

15. *Homocentrica* 2.26, fols. 64ᵛ–65ᵛ, concerning Venus: "What! First we are constrained to construct an epicycle on an epicicyle, then a third on the second?" A theory that relies on epicycles is powerful as long as the objective is to "save" phenomena. Cf. Norwood R. Hanson, "The Mathematical Power of Epicyclical Astonomy," *Isis* 4 (1960), pp. 150–58 (followed by an exchange of letters on epicycles and the Fourier series, *ibid.*, p. 561; and *Isis* 5 (1961), pp. 94–95, 98).

16. Galileo, *Dialogue on the Great World Systems*, trans. (1661) Thomas Salusbury, revised G. de Santillana (Chicago: University of Chicago Press, 1953), p. 350.

17. Cf. below, this chapter, pp. 88–92.

18. Translated by Koyré, *The Astronomical Revolution*, p. 135.

19. In fact, according to the Ptolemaic model, the planets move in the same direction as the epicycles on their deferents. Hence if they were slower than the earth, either Venus or Mercury should move in the opposite direction.

20. Ptolemy, *Almagest* 9.3.

21. Parallax designates the change in the apparent position of a celestial body observed from a point other than the center around which it moves. If the earth revolves around the sun, the position of a star as observed from the earth should appear to shift as the position of the earth moves, unless the distance from the earth to the star is so great that the parallax cannot be observed. Thus Copernicus attributes the absence of parallax in our observations of the fixed stars to "their immense height, which makes even the sphere of the annual

motion, or its reflection, vanish from before our eyes" (OR, p. 22). Such a parallax was not observed until 1838.

22. Concerning discussions about the immobility or motion (and velocity) of the sphere of fixed stars, cf. Michel Lerner, "L'Achille des coperniciens," *Bibliothèque d'humanisme et Renaissance* 42 (1980), pp. 313–27.

23. Cited in Vitruvius, *Architettura* (*dai libri I–VII*), S. Ferri (ed. and trans.) (Rome: Palombi, 1960), p. 57n.

24. In fact Venus requires 7½ months and Mercury 88 days.

25. Stephen Pepper, *The Basis of Criticism in the Arts* (Cambridge, MA: Harvard University Press), 1963, p. 74.

26. Franz and Karl Zeller (eds.), *Nicolaus Kopernikus Gesamtausgabe* (Munich: Oldenbourg, 1949), vol. 2, p. 31.

27. Kepler (GW, vol. 3, p. 23) distinguishes between a Copernicus "who speculates" (dedicatory epistle, Book 1) and a Copernicus "who calculates" (Books 2–6).

28. Cf., for example, *De revolutionibus* 5.25.

29. Koyré, *The Astronomical Revolution*, pp. 150–51.

30. *De revolutionibus* 3.15.

31. Cf. especially Owen Gingerich, "Crisis vs. Aesthetics in the Copernican Revolution," *Vistas in Astronomy* 17 (1974), pp. 85–93.

32. Cf. Derek J. Price, "Contra-Copernicus, in M. Clagett (ed.), *Critical Problems in the History of Science*, pp. 212–14.

33. Cf. Norwood R. Hanson, "The Copernican Disturbance and the Keplerian Revolution," *Journal of the History of Ideas* 22 (1961), p. 175: "Bluntly, there never was a Ptolemaic system of astronomy. Copernicus' achievement was to have invented systematic astronomy."

34. Aristotle, *De caelo* 2.10.291.

35. Elsewhere (*Meteriology* 1.3.341a), Aristotle argues that the moon, while *close* to the earth, moves *slowly*, which explains why it produces no heat. Cf. D. R. Dicks, *Early Greek Astronomy to Aristotle* (Ithaca, NY: Cornell University Press, 1970), p. 215.

36. Cf. P. Marchal, "Discours scientifique et déplacement métaphysique,"

in R. Jongen (ed.), *La Métaphore: Approche pluridisciplinaire* (Brussels: Publications des Facultés Universitaires Saint-Louis, 1980), pp. 99–139. In Copernican studies, Edward Rosen's work is representative of this compulsion to engage in unidimensional interpretation. One series of Rosen's articles, all presented in the form of a question "Was Copernicus X?" and all leading to a negative conclusion, illustrates the will to reduce the speaking subject to an empty "zero" state, the place where science is deployed in all its purity, A formula like "one of the greatest geniuses" or the implication that a Pythagorean or Hermetic influence would detract from his greatness connotes, moreover, a conception of history that, while appealing to objective collections of bare facts, comes close to hagiography: the temptations of Saint Copernicus. See Rosen, "Was Copernicus a Pythagorean"; "Was Copernicus a Hermeticist?" *Minnesota Studies in the Philosophy of Science* 5 (1970), pp. 163–71; and "Was Copernicus a Neoplatonist?" *Journal of the History of Ideas* 44 (1983), pp. 667–69

37. This is the case, for example, with Jean Bernhardt, who considers the comparison to be a commonplace for prefaces, as dictated by the rhetoric of the *captatio benevolentiae*. See Bernhardt, "Légendes coperniciennes et modernité de Copernic," *Revue philosophique de France et de l'étranger* (1980), pp. 151–52. We could oppose this reductive interpretation by arguing that Copernicus, far from making concessions or engaging in a multiplicity of tactical maneuvers, expressed in the same preface his contempt for modern Lactantius and *mataiologoi*, and that he forthrightly expressed his demands on readers: *Mathematica mathematicis* (Mathematics is for mathematicians). Far from defending the thesis that Copernicus' text is unidimensional, Jean Bernhardt recognizes and preaches the necessity of approaching it in the cultural context of the Renaissance.

38. Tycho Brahe, *Opera omnia*, J. L. E. Dreyer (ed.) (Copenhagen, 1913–1929), vol. 6, p. 222 (Letter to Rothmann).

39. In the dialogue *De recte Latini Graecique sermonis pronunciatione*. See Erasmus, *Opera omnia*, ed. J. Clericus (Leyde: Vander Aa, 1703-1706), vol. 4, col. 40). Dürer cites the passage again in the epigraph to *Quattuor institutionum geometricarum libri*.

40. Cf. especially Ernst Kris and Otto Kurz, *Legend, Myth and Magic in the Image of the Artist* (New Haven: Yale University Press, 1979), pp. 38-60.

41. Gauricus, *De sculptura*, ed. Chastel-Klein (Geneva: Droz, 1969), p. 73. The introduction in this edition to the chapter "De symmetria" (pp. 75-91) provides an excellent restatement of the artistic problem.

42. Alberti, *De re aedificatoria* 9.5: "Thus we can establish that beauty is a kind of union and agreement of the parts within the whole to which they belong, according to a determined number, a precise rule, a given situation, just as symmetry [*concinnitas*], that is the absolute and primordial rule of nature, has ordered it.

43. Equicola, *Della Natura d'amore* (Venice: N. de Sabbis, 1531), fol. 78$^{r-v}$: "La bellezza del corpo ricerca che le member siano ben collocate con debiti intervalli, & spatii, ciascuna parte sia con sue sempre commensa, proportione, et conveniente quantità" [The beauty of the body requires members that are well placed with proper intervals and spaces so that each part agrees and is proportionate to the others with suitable quantity].

44. Lomazzo, *Trattato dell'arte* (Milan: Gottardo Pontio, 1585), p. 35 (1.3): "Proportione non è altro ch'una consonanza, & respondenza delle misure delle parte fra se stesse, & col tutto in ogni opera, che si fà, & questa consonanza è da Vitruvio chiamata commodulatione: percioché modulo è detto questa misura che si prende in prima, con laquale & le parti & il tutto si misurano" [Proportion is nothing other than a consonance, and agreement of the parts with each other and with the whole, and this consonance is called by Vitruvius *commodulatione*: for that primary measure against which the parts and whole are measured is called the model.]

45. On Dürer and symmetry, see Erwin Panofsky, *The Life and Art of Albrecht Dürer* (Princeton: Princeton University Press, 1955), pp. 276-78; and *Dürers Kunsttheorie* (Berlin, Reimer, 1915), pp. 141ff.

46. Vitruvius, *De architectura* (Amsterdam: Elzevir, 1649), Appendix *De significatione vocabulorum vitruvianorum*, p. 121.

47. Blaise de Vigénère, *Les Images ou tableaux de platte peinture* (Paris: L'Angelier, 1613), non-paginated preface. The translation is accompanied with

a note: "I am not unaware of what Pliny says concerning this word in Book 34, chapter 8: *Non habet latinum nomen Symmetria*. It is all the more difficult to give an adequate French word, which is why I have rendered and explained it with several."

48. Cited by Y. Bellenger, in Du Bartas, *La Sepmaine*, ed. Bellenger (Paris: Nizet, 1981), vol. 1, p. 179: "Symmetry. The word means proportion." This is the term systematically employed by Perrault in his well-known translation of Vitruvius. In fact, *proportio* points to *symmetria* as the actualization of a norm implicit in the definition of *proportio*. Cf. Panofsky, "History of the Theory of Human Proportions," in *Meaning in the Visual Arts* (Harmondsworth: Penguin Books, 1970), p. 97.

49. Vitruvius, *The Ten Books on Architecture*, trans. M. H. Morgan (New York: Dover Publications, 1960), p. 14 (1.2).

50. *Ibid.*, p. 72 (3.1).

51. Galen, *Placita Hippocratis et Platonis* 5.3. Ernesto Grassi notes that the celebrated formula of the *Wisdom of Solomon*, according to which God created everything by measure, number, and weight (11.20) was written in the second century by a Jew from Alexandria who had direct knowledge of the Pythagorean revival in that city. See Grassi, *Die Theorie des Schönen in der Antike* (Cologne: Du Mont Schauberg, 1962), p. 50.

52. Panofsky, "History of the Theory of Human Proportions," in *Meaning in the Visual Arts* (Garden City, NY: Doubleday, 1955), pp. 62–67.

53. *Ibid.*, pp. 83–87: Villard's method "has but little to do with the measurement of proportions, and from the outset ignores the natural structure of the organism" (p. 83). Geometry is superimposed on the human form "like an independent wire framework" (*ibid.*). Villard "was exclusively concerned with a geometrical schematization of the 'technical' dimensions" (p. 87). Edgar de Bruyne notes that Villard proceeds "not by developing learned speculations on musical and plastic proportions, but by utilizing a long practical and technical tradition." See de Bruyne, *Études d'esthétique médiévale* (Bruges: De Tempel, 1946), vol. 3, p. 255.

54. Dürer, *Quatre livres sur la proportion*, trans. L. Meigret (Paris: Perier,

1557), fol. 102ᵛ.

55. Cited by Panofsky, *Dürers Kunsttheorie*, p. 143.

56. Cited by Panofsky, *ibid.*, p. 143. Cf. also Lomazzo, *Trattato dell'arte* 6.3, p. 285: "Le membre hanno da essere fra di loro simmetre, misurate, e proporzionate armonicamente, si che non si vegga in alcuna figura una testa grande, un petto piccolo, una mano larga, una gamba più lunga dell'altra e simili inconvenienti" [Among themselves, the members must be symmetrical, measured, and harmoniously proportioned, such that one figure does not display a large head, small chest, large hand, one leg longer than the other, and similar disharmonies].

57. Dürer, *Quatre livres sur la proportion*, fol. 103ʳ. The term "monstrous" appeared twice in Meigret's French translation (1557). "This manner of mixing monstrous things" (fol. 101ᵛ); "monstrous portrait and almost a human dream" (fol. 102ᵛ). This motif of monstrosity produced by unsuitable quantitative relationships seems to apply especially to pictorial theory. It is not to be confused with monstrosity based on qualitative incompatibilities (mixing parts belonging to different creatures), which is very widespread in the popular imagination; cf. especially Claude Kappler, *Monstres, prodiges et merveilles à la fin du moyen âge* (Paris: Payot, 1980).

58. Alberti, *De pictura*, p. 74.

59. Alberti, *De re aedificatoria* 7.5; see also 6.2.

60. Piero della Francesca, *De prospettiva pingendi*, ed. N. Fasola (Florence: Sansoni, 1942), p. 128 (my italics).

61. *The Literary Works of Leonardo da Vinci*, J. P. Richter (ed. and trans.) (New York: Phaidon, 1970), nos. 34 and 102.

62. Cited by Panofsky, *La Perspective*, p. 61.

63. Cf. John White, *The Birth and Rebirth of Pictorial Space* (London: Faber & Faber, 1957), p. 122.

64. Cited by Jurgis Baltrusaitis, *Anamorphoses ou magie artificielle des effets merveilleux* (Paris: Perrin, 1969), p. 41.

65. *Ibid.*, p. 117.

66. P. Accolti, *Lo Inganno delgi'occhi* (Florence: 1625), p. 49.

67. Cf. above, Chapter 2, sec. 4 ("The Perspectivist Paradigm").

68. Cited by Giulio Carlo Argan, "The Architecture of Brunelleschi and the Origins of Perspective Theory in the Fifteenth Century," *Journal of the Warburg and Courtauld Institute* 1 (1946), p. 102.

69. *Ibid.*, p. 104. Cf.: "one of the most important innovations in Alberti's treatise [*De pictura*] was perhaps that idea of 'knowledge by comparison' which emerges in opposition to the Scholastic conception of knowledge as *scire per causas.*" Also: "...that principle which for causes, understood as external moving forces, substitutes laws, understood as immanent causes which are produced by the reciprocal co-relation of phenomena."

## CHAPTER IV: SYNECDOCHES

1. *De caelo*, in *Works of Aristotle*, vol. 2, 3.2.300a. On the connections between Aristotle and Copernicus, cf. P. Moreaux, "Copernic et Aristote," in *Platon et Aristote à la Renaissance* (Paris: Vrin, 1976).

2. Du Bartas, *His Divine Weekes and Works*, trans. (1605) J. Sylvester (Gainesville, FL: Scholars' Facsimiles & Reprints, 1965), 4.139-54. Simon Goulart comments, "The Poet was content to touch on this Paradox in a word, without wanting to refute it too exactly...." See Du Bartas, *La Sepmaine*, ed. Y. Bellenger (Paris: Nizet, 1981), vol. 1, p. 154n. K. Reichenberger notes, however: "Das von Du Bartas verwendete Bild setzt genaue Kenntnis der kopernikanischen Theorien, wenn nicht gar Lektre des Werkes selbst voraus" [The image employed by Du Bartas presumes an accurate knowledge of Copernican theory, although not at all a reading of the work itself.]. See Reichenberger, *Themen und Quellen der Sepmaine* (Munich: Hueber, 1962), p. 154.

3. Brahe, *Opera omnia*, vol. 7, p. 219. Cf. Alexandre Koyré, *Galileo Studies*, trans. John Mepham (Atlantic Highlands, NJ: Humanities Press, 1978), pp. 141-53.

4. On the distinction between the particularizing synecdoche (the part for the whole, the particular for the general) and the generalizing synecdoche (the whole for that part, the general for the particular), cf. Groupe *Mu, Rhétorique générale* (Paris: Larousse, 1970), pp. 102-04. The Copernican synecdoche is par-

ticularizing, since it substitutes the "less" (the parts of the universe) for the "more" (the totality of the universe) as a physical whole.

5. Aristotle, *De caelo* 4.3.310b.

6. *Physica*, in *Works of Aristotle*, trans. R. P. Hardie and R. K. Gaye (Oxford: Clarendon Press, 1930), vol. 2, 208b.

7. Maurice Clavelin, *La Philosophie naturelle de Galilée* (Paris: A. Colin, 1968), p. 47.

8. Leading to Edward Grant's remark, "Late Medieval Thought, Copernicus and the Scientific Revolution," *Journal of the History of Ideas* 23 (1962), p. 215: "Copernicus makes one more important innovation. He seems to subordinate physics to astronomy, thus reversing an ancient and medieval tradition."

9. The essential texts (with commentary) are collected in Marshall Clagett, *The Science of Mechanics in the Middle Ages* (London: Oxford University Press, 1959), p. 614. Concerning Buridan's influence in Poland, see Markowski, *Buridanizm w polsce w okresie przedkopernickanskim* (Wroclaw: Presses de l'Académie de Pologne, 1971). We should note in passing that Henri Busson attributes to Copernicus' influence (ROC, p. 16) Rabelais' use in the *Le Cinquième livre* (ch. 26) of the impressions of a man on a moving boat to illustrate the optical relativity of motion. See Busson, *Le Rationalisme en France au XVIᵉ siècle* (Paris: Vrin, 1957), p. 256. But many other sources are possible (Buridan, Oresme, Nicholas de Cusa); see Raymond Klibansky, "Copernic et Nicolas de Cuse," in *Léonard de Vinci et l'expérience scientifique au XVIᵛ siècle* (Paris: Presses Universitaires de France, 1953), pp. 225–35.

10. Pierre Duhem, *Études sur Léonard de Vinci* (Paris: Hermann, 1906), vol. 2, pp. 301ff.

11. Cf. Alexandre Koyré, *From the Closed World to the Infinite Universe* (Baltimore: Johns Hopkins University Press, 1957), p. 19. On the connections between Nicholas de Cusa and Copernicus, cf. also Klibansky, "Copernic et Nicolas de Cuse."

12. Cf. Pierre Kerszberg, "La cosmologie de Copernic et les origines de la physique mathématique," *Revue d'histoire des sciences* 34 (1981), p. 23. See also the *Timaeus* (34a–b and 36e) and the *Laws* (897e–898b).

13. Z. Horsky, "Le rôle du platonisme dans l'origine de la cosmologie moderne," *Organon* 4 (1967), pp. 47–54; and "Intervention," in *Avant, avec, après Copernic*, Centre International de Synthèse (Paris: Blanchard, 1975), pp. 97–98.

14. On this point, cf. Paul Kristeller, *The Philosophy of Marsilio Ficino* (Gloucester, MA: Smith, 1964), ch. 10. As Kerszberg notes ("La cosmologie," p. 5), "questions of physics developed within the framework of Neoplatonic philosophy were...often carried over into the framework of Aristotelian language." I should add that Copernican physics has also been integrated into a geocentric model of the universe; this is what takes place with Patrizi (*Pancosmia* 17.102.2, in his *Nova de universis philosophia*).

15. Ficino, *Théologie platonicienne*, Raymond Marcel (ed. and trans.) (Paris: Les Belles Lettres, 1964), vol. 1, p. 160 (4.1).

16. *Ibid.*, p. 151.

17. We can note in this respect, the numerous inferences from terrestrial to celestial order that Ficino sprinkled throughout this chapter, especially pp. 150 and 159.

18. *Ibid.*, p. 163. In fact Aristotle's position is not very clear on this point. Cf. the introduction to the edition by William Guthrie, *On the Heavens* (Cambridge, MA: Harvard University Press, 1954), pp. xxix–xxxvi. But the important thing here is Ficino's trenchant interpretation.

19. Cf. Edgar Zilsel, "Copernicus and Mechanics," *Journal of the History of Ideas* 1 (1940), pp. 113–18.

20. Ficino, *Theologica platonica*, vol. 1, p. 144.

21. *Ibid.*, p. 160. We should also note that Copernican physics brings into play a network of qualitative affinities arising from what Michel Foucault has called the *episteme* of likeness. See Foucault, *The Order of Things* (New York: Pantheon, 1971), ch. 2. In Ficino, we find the four figures of similarity recorded by Foucault. Proximity is the sign of relationship, of participation in a common nature (the terrestrial, for example): *suitability*. Distant objects conform by their tendency to form a sphere: *analogy* between celestial bodies and a drop of water (OR, p. 8). Terms like *appetere*, *appententia*, *cupere*, and *contendere* suggest in a repetitive manner a common *emulation* of the desire to return to per-

THE POETIC STRUCTURE OF THE WORLD

fect form. Finally, if the moving earth pulls the nearby air with it, this indicates a *sympathetic* relation (OR, p. 16). And so, the process of geometrization does not exclude another form of rationality but rather reinterprets it: the qualitative must submit to a geometric final cause, namely the realization of spherical unity.

22. Ficino, *Theologica platonica*, vol. 1, p. 146. Cf. also p. 160.

23. Koyré, *The Astronomical Revolution*, p. 58.

24. Cf. Erwin Panofsky, *The Codex Huyghens and Leonardo da Vinci* (London: Warburg Institute, 1949).

25. Leonardo da Vinci, *The Literary Works of Leonardo da Vinci*, vol. 1, p. 273, nos. 394–96. As we know, growth was considered to have the same nature as motion, both being forms of *change*. For Leonardo, on the other hand, reflections on motion established a tight link between painting and philosophy (vol. 1, p. 37, no. 9).

26. Severo Sarduy, *Baroco*, p. 48. Concerning the circle for Raphael and Copernicus, cf. also *The Nature of Scientific Discovery*, O. Gingerich (ed.) (Washington, D.C.: Smithsonian Institution Press, 1975), p. 473.

27. Cf. H. Damisch, *Théorie du nuage* (Paris: Éd. du Seuil, 1972), pp. 239–40. Aristotelian language, however, did not preclude Platonic overtones. Cf. André Chastel, *Marsile Ficin et l'Art* (Geneva: Librairie Droz, 1975), pp. 93–94; and *Art et humanisme à Florence* (Paris: Presses Universitaires de France, 1961), pp. 308–13.

28. *The Literary Works of Leonardo da Vinci*, vol. 2, p. 109, no. 858: "Come la terra non è nel mezzo del cerchio del solen, ne nel mezzo del mondo, ma è ben nel mezzo de' sua elementi conpagni et uniti con lei" [The earth is not in the center of the sun's orbit nor at the center of the universe, but in the center of its companion elements, and united with them].

29. *Ibid.*, vol. 2, p. 130, no. 902: "Tal luna è vestita de' sua elementi, cioè acqua, aria, e foco, e cosi in sé, per sé si sostenga in quello spatio, come fa la nostra terra coi sua elementi in quest' altro spatio; e...tal offitio faccino le cose gravi ne' sua elementi qual fanno l'altre cose gravi nelli elementi nostri" [The moon is surrounded by its own elements: that is, of itself and by itself, suspended

in that part of space, as our earth with its elements is in this part of space; and...heavy bodies act in the midst of its elements just as other heavy bodies do in ours]. For Plato also, the stars are composed of four elements (*Timaeus* 31b–33b), of which fire is predominant (*Timaeus* 40a).

30. *Ibid.*, vol. 2, p. 110, no. 861: "Movasi la terra da che parte si voglia, mai la superfitie dell'acqua uscirà fori della sua spera, ma sempre sarà equidistante dai centro del mondo. Dato che la terra si rimovessi dal centro del mondo, che farebbe l'acqua? Resterebbe intorno a esso centro con equal grossezza, ma minore diametro, che quando ella avea la terra in corpo" [Let the earth turn on which side it may, the surface of the waters will never move from its spherical form, but will always remain equidistant from the center of the globe. Granting that the earth might be removed from the center of the globe, what would happen to the water? It would remain in a sphere round that center equally thick, but the sphere would have a smaller diameter than when it enclosed the earth.] Cf. also *ibid.*, vol. 2, p. 147, no. 936. In no. 861, Richter translates "mondo" as globe, in no. 936 as "world."

31. Alexandre Koyré, *Études d'histoire de la pensée scientifique*, p. 170.

32. Giorgio de Santillana, "The Role of Art in the Scientific Renaissance," in M. Claggett (ed.), *Critical Problems in the History of Science*, pp. 59–60.

33. OR, p. 18. On the relationship for Copernicus between homogeneous space and the cosmos, see especially Kerzberg, "La Cosmologie."

34. Argan, "The Architecture of Brunelleschi and the Origins of Perspective Theory in the Fifteenth Century," pp. 115–21.

35. Cf. Klein's commentary in Gauricus, *De sculptura*, Chastel-Klein (ed.) (Geneva: Droz, 1969), pp. 177–78; Klein, *La Forme et l'intelligible* (Paris: Gallimard, 1970), pp. 237–77; and Damisch, *Théorie du nuage*.

36. Gauricus, *De sculptura*, p. 197.

37. Cf. Klein, *La Forme et l'intelligible*. This poetic and rhetorical space will become dominant in Mannerism. Painters will have a tendency to conceal the grid, which provides the basis of geometric constructs, making the picture a relationship of actors, a relationship that creates rather than occupies space. Cf. Dagobert Frey, *Manierismus als europäische Stilerscheinung* (Stuttgart: Kohl-

hammer, 1964), p. 34.

38. Gauricus, *De sculptura*, p. 197.

39. Lomazzo, *Idea*, cited in Klein, *La Forme et l'intelligible*, p. 245.

40. Gauricus, *De sculptura*, p. 197.

41. Ptolemy, *Almagest* 3.3, cited in Pierre Duhem, *Le Système du monde*, vol. 1, p. 487.

42. Cf. Proclus, *Hypothesis astronomicarum positionum* 7.

43. Fracastoro, *Homocentrica*, fol. 1ʳ: "They understand those divine bodies in an erroneous and, to some degree, impious fashion, giving them positions and forms altogether unsuitable to heaven." Cf. also 2.26.

44. The theory of the *Commentariolus* differs in two ways from that of the *De revolutionibus*. In the former, the system of representation is "concentric and "bi-epicyclical": the center of the deferents coincides with the center of the universe and a second epicycle is grafted on the first. In the latter, it is eccentric and epicyclical: the deferent is eccentric, and there is only one epicycle. Moreover, the apsides are stationary in the *Commentariolus* (as they are for Ptolemy), but mobile in the *De revolutionibus*.

45. Cf. Owen Gingerich, "The Role of Erasmus Rheinhold and the Prutenic Tables in the Dissemination of the Copernican Theory," *Colloquia Copernicana* 2 (Warsaw: Ossolineum, 1973). Copernicus himself entitled Book 1, ch. 4 in the following emphatic manner: "The motion of the heavenly bodies is uniform, eternal, and circular or compounded of circular motion" (OR, p. 10).

46. Bernard Meyer, "Sous les pavés, la plage: Autour de la synecdoque du tout," *Poétique* 62 (1985), p. 195.

47. Brahe, *Opera omnia*, vol. 1, p. 149. Cf. also Galileo, *Opere* (Florence: Barbera, 1968), vol. 7, p. 369: "Sono in Tolomeo le infermità, e nel Copernico i medicamenti loro. E prima no chiameranno tutte le sette di i filosofi grande sconvenevolezza [impropriety] che un corpo naturalmente mobile in giro si muova irregolarmente sopra un'altro punto? e pur di tali movimenti difformi sono nella fabbrica di Tolomeo, ma nel Copernico tutti sono equabili intorno al proprio centro" [In Ptolemy there are maladies, and in Copernicus remedies for those maladies. And did not all seven of the great philosophers call it an impro-

priety that a body that naturally moves in a circuit should move irregularly around another point? And likewise such deformed motions are in Ptolemy's structure, but in Copernicus all bodies move uniformly around their own center.]

48. Ficino, *Theologica platonica* 3.2, vol. 1, p. 140.

49. Otto Neugebauer, "On the Planetary Theory of Copernicus," *Vistas in Astronomy* 10 (1968), pp. 89-2¼-113; and N. Swerdlow, "The Derivation and First Draft of Copernicus' Planetary Theory," *Proceedings of the American Philosophical Society* 117 (1973), pp. 423-512

50. Cf. especially Derek J. Price, "Contra-Copernicus," in Clagett (ed.), *Critical Problems in the History of Science*, pp. 204-05.

51. A. G. Molland, "Ancestors of Physics," *History of Science* 14 (1976), p. 63; cf. Rose, *The Italian Renaissance of Mathematics*, pp. 2-3.


CHAPTER V: METAPHOR OF THE CENTER

1. Concerning the numerous antecedents for the motifs appearing in Copernicus' panegyric to the sun, cf. F. and K. Zeller (eds.), *Nicolaus Kopernikus Gesamtausgabe*, vol. 2, pp. 441-44. Concerning Copernicus' reference to Sophocles (*Oedipus at Colonus*, 1.869, not *Electra*), cf. Edward Rosen, "Copernicus' Quotation from Sophocles," in S. Prete (ed.), *Didascaliae: Studies in Honor of A. M. Albareda* (New York: Rosenthal, 1961), pp. 369-79.

2. Commentary by Alexandre Koyré in Copernicus, *Des Revolutions des orbes célestes*, trans. Koyré (Paris: Alcan, 1934), p. 147. Cf. also p. 23: "Ancient traditions, the tradition of the metaphysics of light (which throughout the Middle Ages accompanied and supported the study of geometric optics), Platonic and Neoplatonic reminiscences (the visible sun representing the invisible sun), these alone can, in my judgment, explain the emotion, the lyricism which takes hold of Copernicus when he speaks of the sun. He worships it, almost deifies it. The splendid light shining on the universe becomes the ontological center and thereby the geometric center of the Universe." Koyré takes up the same theme in *The Astronomical Revolution*, p. 65.

3. Frances Yates, *Giordano Bruno and the Hermetic Tradition* (Chicago: University of Chicago Press, 1964), p. 153.

4. *Ibid.*, p. 452: "It may be illuminating to view the scientific revolution as in two phases, the first phase consisting of an animistic universe operated by magic, the second phase of a mathematical universe operated by mechanics." Cf. also Yates, "The Hermetic Tradition in Renaissance Science," in C. S. Singleton (ed.), *Art, Science and History in the Renaissance* (Baltimore: Johns Hopkins University Press, 1970), p. 271: "The emergence of modern science should perhaps be regarded as proceeding in two phases, the first being the Hermetic or magical phase of the Renaissance with its basis in an animist philosophy, the second being the development in the seventeenth century of the first or classical period of modern science."

5. Edward Rosen, "Was Copernicus a Hermetist?" pp. 168–69. Other arguments by Rosen are less convincing. It is true that the expression "visible god" did not literally appear in the revelations of Hermes Trismegistus. But *Asclepius* and *Poimandres* both call the Sun a *god*: *deus vivantium, secundus deus, deus praestantissimus* [living god, second god, most present god] (*Poimandres* 11, *Asclepius* 11), etc. The basis for the metaphor appears repeatedly. As for the adjective *visible*, it sums up such an elaboration as this:

> Finally, *since you want to see God, raise your eyes* to the other stars; who eternally preserves their order? Any order assuredly is defined by the limits of a number, of a location. The Sun is the most illustrious god of the celestial gods. The other heavenly gods obey the Sun like a prince and a king. The Sun [is] so great, vaster than the earth and sea, nevertheless permits innumerable minor stars to turn around him. (*Poimandres*, 11)

This passage also characterizes the entire heavens as a visible symbol of divinity; Copernicus does the same in his introduction to Book 1 (OR, pp. 7–8). Rosen considers a spelling error in the manuscript of *De revolutionibus* ("Trimegistus" for "Trismegistus") as an indicator of Copernicus' lack of familiarity with Hermeticism. But in the same chapter, Copernicus makes numerical errors concerning the periods of Venus and Mars' orbits. Zeller, *Nicolaus Kopernikus Gesamtausgabe*, vol. 2, p. 429, includes an entire page of minor errors of the same sort, including *the* major error *immobilitate telluris* [earth's immobility] for *mobilitate telluris* [earth's mobility]. What conclusions would we draw if we judged these with the same severity? It is worth noting that the importance of Hermeticism for Copernicus was recognized by Mersenne in *Quaestiones in Genesim*, col. 892.

6. Jean Bernhardt, "Légendes coperniciennes et modernité de Copernic," p. 149.

7. The sun also did not play a dynamic role in the geocentric universe, where the same metaphors were current.

8. For Brahe, cf. Robert S. Westman, "Three Responses to the Copernican Theory," in R. S. Westman (ed.), *The Copernican Achievement*, p. 319. Kepler's "sun worship" is omnipresent in his work (cf. especially GW, vol. 13, p. 35). For Mersenne, see *Quaestiones in Genesim*, col. 892n.5. By the number of reactions that he provoked, the passage belongs to what Michael Riffaterre calls the "arch-reader," that is the collection of "points where the attention of *n* readers, critics, analysts, etc. converge." See Riffaterre, *Essais de stylistique structurale* (Paris: Flammarion, 1971), p. 47.

9. Garin, *Rinascite et rivoluzioni*, pp. 255–81. On solar symbolism in the Middle Ages, cf. Helen F. Dunbar, *Symbolism in Medieval Thought* (New Haven: Yale University Press, 1929).

10. Ficino, *Theologica platonica*, 18.3. Cf. also *De vita libri tres* 3.1, and the brief text entitled *Orphica comparatio solis ad Deum*, in *Opera* (Paris: Bechet, 1641), pp. 797-98. *De sole* was known in Poland by the end of the fifteenth century: cf. E. Rubka, "The Influence of the Cracow Intellectual Climate at the End of the Fifteenth Century upon the Origin of the Heliocentric System," *Vistas in Astronomy* 9 (1967), p. 165.

11. Cf. Garin, *Rinascite e rivoluzioni*, p. 294. The same biblical verse also appears in the passage cited from Ficino's *Theologica platonica* as well as in his *De sole*. On the solar theme in Pico della Mirandola, cf. also Raymond B. Waddington, "The Sun at the Center: Structure as Meaning in Pico della Mirandola's *Heptaplus*," *Journal for Medieval and Renaissance Studies* 3 (1973), pp. 69-86.

12. It is unnecessary to recall examples here. Cf. the collection of studies in *Le Soleil à la Renaissance* (Brussels: Presses Universitaires de Bruxelles, 1965).

13. Galileo, *Opere* (Florence: Barbera, 1968), vol. 1, p. 52. Cf. William Shea, "Le Copernicanisme de Galilée," in *Avant, avec, après Copernic*, pp. 213–18.

14. Cf. M. Righetti, *Manuale di storia liturgica* (Milan: Ancora, 1959–1969), vol. 1, pp. 474ff. The solar monstrance will triumph in Baroque art. Cf. also Emile Mâle, *L'Art religieux après le concile de Trente* (Paris: Colin, 1936); and C. Constantini, *Dio nascosto: Plendori di fede e d'arte nella S. Euraristia* (Rome, 1944).

15. *De caelo*, in *Works of Aristotle*, vol. 2, 4.1.308a.

16. The paradigmatic example is of course Dante's *Inferno*. Cf. also Garin, *Rinascite e rivoluzioni*, p. 291n.

17. Ficino, *De sole*, ch. 13 (entitled "Solem non esse adorandum tanquam rerum omnium authorem" [The sun is not to be worshiped as if it were the author of all things]).

18. Concerning this frontispiece, cf. Jean Seznec, *The Survival of the Pagan Gods* (New York: Pantheon Books, 1953), pp. 140–43; Edgar Wind, *Pagan Mysteries in the Renaissance* (London: Faber & Faber, 1958), pp. 46, 112–13; and Stanislaw Mossakowski, "The Symbolic Meaning of Copernicus' Seal," *Journal of the History of Ideas* 34 (1973), pp. 451–60.

19. Cf. Ficino, *Theologica platonica* 3.2, vol. 1, pp. 137ff.

20. Paul Kristeller, *Huit Philosophes de la Renaissance italienne* (Geneva: Droz, 1975), p. 48.

21. Mircea Eliade, *Images and Symbols*, trans. P. Mairet (New York: Sheed & Ward, 1961), p. 39.

22. Ficino, *De lumine* (*Opera*, vol. 1, p. 999): "But neither in this book, nor in the preceding, did I want to reflect on the rather minor questions of the Mathematicians concerning the Sun or of Light, [for they are] often less useful than they assuredly are difficult. [I preferred] rather, to make use of some of their examples and their comparisons to proceed step by step in the quest of the intentions, rules and divine mysteries of the spirit that we must contemplate."

23. *De caelo* (2.13.293a–293b), where Aristotle contrasts his conception of the geometric center ("what is defined," the "matter" rather than "what de-

fines," the "essence") and the Pythagorean conception, which places fire at the center of the universe, considered by Pythagoreans "the most precious place."

24. *Le Pimandre*..., (Bordeaux; Millanges, 1579), p. 701. On François de Foix's role in spreading the writings of Hermes Trismegistus, cf. Raymond Marcel, "La Fortune d'Hermes Trismégiste à la Renaissance," in *L'Humanisme français au début de la Renaissance* (Paris: Vrin, 1973), pp. 137–54. A. Rosselli's long commentary (*Divinus Pymander*..., 6 vols., Cologne, 1630), while taking up de Foix's translation and celebrating the sun at length, does not say a word about Copernican theory.

25. Note that God's *repose*, his "immobile eternity, in which turbulent time ebbs and has its beginning," was strongly emphasized in *Asclepius* 11. In *De sole*, Ficino noted that by its motion the sun differs from what it represents: "Since repose, as a principle, master, and final end of motion, is more perfect than any motion, certainly God Himself, principle and master of all things, cannot be in motion. The sun, on the other hand, is incessantly in motion" (*Opera*, vol. 1, p. 999).

26. Cited by Rudolf Wittkower, *Architectural Principles in the Age of Humanism* (New York: Norton, 1971), p. 12.

27. Ptolemy, *Almagest* 1.1.

28. Cited by Wittkower, *Architectural Principles*, p. 12. Raphael, known for his quest for circular unity (cf. above, Chapter 4, p. 115) as well as for his vertical symbols associated with circular forms (cf. above, p. 133, concerning *La Disputa*), also painted in *The Marriage of the Virgin* a circular temple within a circular space created by the arrangement of the characters.

29. *Ibid.*, pp. 19-20

30. Klein, *La Forme et l'intelligible*, p. 314. Cf. also, Roland Le Moll, "La Ville idéale," *Bibliothèque d'Humanisme et Renaissance* 33 (1971), pp. 689–702.

31. Klein, *La Forme et l'intelligible*, pp. 318–20. The center is always favored, but sometimes as an empty space (in liberal utopias), sometimes as the seat of religious and/or political power.

32. Cf. Jean Wirth, "La naissance du concept de *croyance*," *Bibliothèque d'humanisme et Renaissance* 45 (1983), pp. 7–58.

33. Erasmus, *Opera Omnia*, ed. J. Clericus (Leyde: Vander Aa, 1703-1706), vol. 5, col. 141 (*Paracelsis*)

34. Erasmus, *Correspondence* (Toronto: University of Toronto Press, 1974), vol. 5, p. 218 (Letter 1523).

35. Theresa of Avila, *Moradas*, T. N. Tomas (ed.) (Madrid: Espasa Calpe, 1947), pp. 6-7 n. (1.1).

36. On Theresa of Avila's tendency to allegorize, cf. Helmut Hatzfeld, *Estudios literarios sobre mistica española* (Madrid: Gredos, 1968), pp. 207-09, 218.

37. *The Collected Works of St. John of the Cross*, trans. K. Kavanaugh and O. Rodgriguez (Camden, NJ: Nelson, 1964), p. 583.

38. Pierre de Bérulle, "Second discours," *Discours de l'estat et des grandeurs de Jésus*, in *Oeuvres* (Paris: Léonard, 1665), pp. 115-16.

39. Henri Brémond, *Histoire littéraire du sentiment religieux en France* (Paris: Colin, 1968), vol. 2, pp. 26ff.

40. Dagens, *Bérulle et les origines de la restauration catholique* (Paris: Desclée de Brouwer, 1952), p. 22: "it is probable that this passage by François de Foix inspired Bérulle. Bérulle enthusiastically cites Pimandros, the Egyptians, and also Orpheus. We must avoid modernizing him."

41. On this theme in Pico della Mirandola, cf. Waddington, "The Sun at the Center," pp. 75-76, 78.

42. Koyré, *The Astronomical Revolution*, p. 59-65.

43. Clémence Ramnoux, "Héliocentrisme et christocentrisme (sur un texte du cardinal de Bérulle)," in *Le Soleil à la Renaissance*, p. 458.

44. Carolo Borromeo, *Instructiones fabricae et supellectilis ecclesiasticae 2* ("De ecclesiae forma"), in P. Barocchi (ed.), *Trattati d'arte del Cinquecento* (Bari: Laterza, 1962), vol. 3, pp., 9-10.

45. Giovanni Battista Riccioli, *Almagestum novum* (Bononiae: Benatius, 1661), vol. 2, p. 469.

46. In Jean Rousset, *Anthologie de la poésie baroque française* (Paris: Colin, 1961), vol. 2, p. 278.

47. The characteristic traits of the analogical model covered here are those catalogued by Dedre Gentner, "Are Scientific Analogies Metaphors?" in D. S.

Miall (ed.), *Metaphor: Problems and Perspectives* (Atlantic Highlands, NJ: Humanities Press, 1982), pp. 106–32.

CHAPTER VI: FROM COPERNICUS TO KEPLER

1. In addition to the studies cited in the following notes, cf. Dorothy Stimson, *The Gradual Acceptance of the Copernican Theory of the Universe* (New York: Baker & Taylor, 1917); Ernst Zinner, *Entstehung und Ausbreitung der copernicanischen Lehre*; Jerzy Dobrzycki (ed.), *The Reception of Copernicus' Heliocentric Theory* (Boston: Reidel, 1972); and Lynn Thorndike, *History of Magic and Experimental Science* (New York: Columbia University Press, 1941), vols. 5ff.

2. Koyré, *Galileo Studies*, pp. 132–35.

3. Cf. Especially Klaus Scholder, *Ursprünge und Probleme der Bibelkritik im 17. Jahrhundert* (Munich: Kaiser, 1966), pp. 56–78 ("Kopernikus und die Folge"); and Jean Stengers, "Les Premiers coperniciens et la Bible," *Revue belge de philologie et d'histoire* 62 (1984), pp. 703–19.

4. Cf. Montaigne, "Apology for Raymond Sebond," p. 429: "and in our day Copernicus has grounded this doctrine so well that he uses it very systematically for all astronomical deductions. What are we to get out of that, unless that we should not bother which of the two is so? And who knows whether a third opinion, a thousand years from now, will not overthrow the preceding two?"

5. Cf. E. H. Waterbolk, "The 'Reception' of Copernicus' Teachings by Gemma Frisius (1508–1555)," *Lias* 1 (1974), pp. 225–41.

6. Cf. Robert S. Westman, "Three Responses to the Copernican Theory," in R. S. Westman (ed.), *The Copernican Achievement*, p. 293.

7. François de Foix, *Le Pimandre...*, p. 701. On the first Copernican echoes in France, cf. Jean Plattard, "Le Système de Copernic dans la littérature française au XVIᵉ siècle," *Revue du seizième siècle* 1 (1913), pp. 220–37; Henri Busson, *Le Rationalisme en France au XVIᵉ siècle* (Paris: Vrin, 1957), pp. 257–58; Beverly S. Ridgely, "Mellin de Saint-Gelais and the First Vernacular Reference to the Copernican System in France," *Journal of the History of Ideas* 23 (1962), pp. 107–16 (with complementary bibliography); and Thorndike, *History of Magic*

and Experimental Science, vol. 6, *passim*.

8. Mulerus, *Institutionum astonomicarum libri duo* (Groningue: Saffius, 1616), pp. 14–15.

9. Cf. Westman, "Three Responses," pp. 286–87.

10. Cf. Westman, "The Melanchthon Circle, Rheticus, and the Wittenberg Interpretation of the Copernican Theory," *Isis* 66 (1975), pp. 165–93; a different version of this article appeared as "The Wittenberg Interpretation of the Copernican Theory," in O. Gingerich (ed.), *The Nature of Scientific Discovery*, pp. 393–429, with discussion pp. 430–57. On Praetorius, cf. the article cited in the previous note.

11. Westman, "Three Responses," pp. 289–305.

12. Dodoens, *Cosmographica in astronomica et geographical isagoge*, facsimile ed. (1548) with introduction by A. Louis (Nieuwkoop: De Graaf, 1964); and *De sphaera, sive de astronomiae et geographiae principiis isagoge* (Anvers: Plantin, 1584). Contrary to what Louis says (p. 36), there are numerous differences between the two editions. In the first, Copernicus is only mentioned once (2.14); in the second, he is mentioned five times (Preface, 1.7, 2.14, 3.7, 4.17).

13. Dodoens, *De sphaera*, p. 4. "In effect, [my work] leads [the reader] by the hand toward this man's books on the revolutions."

14. *Ibid.*, p. 62 (3.7). These appear in a one-and-a-half page addition to the text of the first edition. Dodoens refers to Book 4 of the *De revolutionibus* (chs. 17ff.)

15. Giordano Bruno, *De infinito* 1, in G. Gentile and G. Aquilecchia (eds.), *Dialoghi italiani* (Florence: Sansoni, 1958), p. 380.

16. Bruno, *La Cena de le ceneri* 1, in *Dialoghi italiani*, p. 28.

17. Bruno, *La Cena de le ceneri* 5, in *Dialoghi italiani*, pp. 160ff.

18. Bruno, *De immenso* 3.10 and 3.8, in *Opera Latini* 1.1, p. 397, and 1.2, p. 145. For the passage in *De Infinito* 3, accepting Copernican "symmetry," cf. *Dialoghi italiani*, p. 437. For analysis and graphic illustration, see Westman, "Magical Reform and Astronomical Reform," in R. S. Westman and J. E. McGuire, *Hermeticism and the Scientific Revolution* (Los Angeles: University of California, 1977), pp. 32–34.

19. Cf. Westman, "Three Responses, pp. 313–21; and K. P. Moesjgaard, "Copernican Influence on Tycho Brahe," in Dobrzycki (ed.), *The Reception*, pp. 31–35.

20. Kuhn, *The Copernican Revolution*, p. 159.

21. Tycho Brahe, *Opera omnia*, vol. 6, pp. 221–22.

22. Froidmont, *Vesta sive Ant-Aristarchus Vindex* (Anvers: Plantin-Moretus, 1634), p. 151. Froidmont directs the same criticism toward Brahe in *Ant-Aristarchus sive orbis terrae immobilis liber unicus* (Anvers: Plantin-Moretus, 1631), ch. 12.

23. Giovanni Battista Riccioli, *Almagestum novum* (Bononiae: Benatius, 1661), vol. 2, p. 459.

24. Cited in H. Hugonard-Roche, E. Rosen, and J.-P. Verdet (eds.), *Introductions à l'astronomie de Copernic* (Paris: Blanchard, 1975), p. 131.

25. On Maestlin and Copernicus, cf. R. S. Westman, "Michael Maestlin's Adoption of the Copernican Theory," *Colloquia Copernicana* 4.

26. GW, vol. 1, pp. 435–36. On the debate concerning this velocity, cf. Michel Lerner, "L'Achille des coperniciens," *Bibliothèque d'Humanisme et Renaissance* 42 (1980), pp. 313–27.

27. Cf. especially the third day of the *Dialogue on the Great World Systems*, pp. 335ff.

28. Cf. H. Lausberg, *Handbuch der literarischen Rhetorik*, §258, pp. 1057–59.

29. Cf., for example, Calvin's commentary on *Genesis, Commentarii in quique libros Misis, Opera*, vol. 23, cols. 22–23 (= *Corpus reformatorum*, vol. 51). On this question, see Stengers, "Le Premiers coperniciens et la Bible."

30. Cf. Amos Funkenstein, "The Dialectical Preparation for Scientific Revolution," in Westman (ed.), *The Copernican Achievement*, pp. 184–85. In this sense, the Copernican universe can also appear to prefigure Descartes' "evil genius": cf. Hans Blumenberg, *Kopernikus im Selbstverständnis der Neuzeit* (Mayence: Akademie der Wissenschaften und der Literatur, 1965), pp. 15–16.

31. Cited by Tibor Klaniczay, *Renaissance und Manierismus* (Berlin: Akademie Verlag, 1977), p. 165. On the Mannerist split between form and image, cf.

also Giulio Carlo Argan, *L'Europe des capitales* (Geneva: Skira, 1964), p. 15; and Robert Klein, *La Forme et l'intelligible*, ch. 5.

32. Cf. Gingerich, (ed.), *The Nature of Scientific Discovery*, p. 450.

33. Cf. Arnold Hauser, *Mannerism* (New York: Knopf, 1964), vol. 1, p. 21.

34. Erwin Panofsky, *Idea* (New York: Harper & Row, 1968), p. 82.

35. *Ibid.*, p. 83.

36. *Ibid.*

37. Cf. Koyré, *The Astronomical Revolution*, pp. 159-71. It is true that subsequently Kepler was scrupulously faithful to the facts, but this submission did not suppress his anagogical objective.


CHAPTER VII: THE SEMIOSIS OF THE WORLD

1. Cf. Kepler's letter to Georg Brennger, October 4, 1607 (GW, vol. 16, p. 54): "I provide in effect a single celestial philosophy or physics in place of Celestial Theology or Aristotle's Metaphysics...." It is worth recalling the complete title of the *Astronomia nova: New Astronomy by Causes, or Celestial Physics....*

2. GW, vol. 1, pp. 23-24. On the importance for Kepler of the principle of sufficient reason, cf. Alexandre Koyré, *From the Closed World to the Infinite Universe* (Baltimore: Johns Hopkins University Press, 1957), ch. 3 (especially p. 78).

3. All these considerations occupy chs. 4-6 of *De stella nova*. Earlier, in chs. 11-12 of the *Mysterium cosmographicum*, Kepler had attempted to justify the division of the zodiac into twelve parts. He abandoned this position for the reasons given in *De stella nova* and in the notes accompanying the aforementioned chapters of the second edition of the *Mysterium cosmographicum*.

4. GW, vol. 1, p. 285. This critical attitude did not prevent Kepler from taking an interest in the similarities of words from different languages – *Jubal-Apollo, Tubalcan-Vulcan* (GW, vol. 6, p. 94) – but he did not expect to unlock profound secrets by means of such an examination.

5. GW, vol. 6, pp. 386 and 396. On this point, cf. Yates, *Giordano Bruno*, pp. 441-44.

6. Translated into English by C. Hardie as *The Six-Cornered Snowflake* (Oxford: Clarendon Press, 1966).

7. "We see that God created the bodies of the universe according to a determinate number. But number is a property of quantity.... That is why, if the universe is founded on numerical measure, the measure must be of quantities" (GW, vol. 13, p. 35). Cf. also GW, vol. 8, p. 60.

8. Aquinas, *Summa theologica* 1.15.1.

9. Cf. Michael Landmann, "Die Weltschöpfung im *Timaios* und in der *Genesis*," in *Ursprungsbild und Schopferstat. Zum platonisch-biblischen Gespräch*, Sammlung Dialog 8 (1966), p. 142.

10. On the notion of *disegno*, cf. P. Barocchi (ed.), *Scritti d'arte del Cinquecento* (Milan: Ricciardi, 1963), vol. 2, pp. 1897–2118 (especially Zuccari, p. 2066). Supplement with Erwin Panofsky's commentary in *Idea*, ch 5. Gustav Hocke summarizes the priority of the *disegno interno* as follows: "A work of art is the result of an artist's idea, not a copy of nature." See Hocke, *Die Welt als Labyrinth* (Hamburg: Rowohlt, 1957), p. 44.

11. Cited by Panofsky, *Idea*, pp. 229–30.

12. *Ibid.*, p. 186.

13. J.-M. Wagner, "Théorie de l'image et pratique iconologique," *Baroque* 9-10 (1980), p. 63.

14. *Ibid.*

15. GW, vol. 2, p. 16: "the universal theatre was organized such that it contains sufficient signs, capable not only of inviting *men's spirits, simulacra of God's spirit*, to contemplate divine works, which can permit them to measure God's goodness, but also of helping them to explore these works fully" (my italics).

16. We must understand that "since God created everything throughout the universe according to the norms of quantity, intelligence was also given to man so that he would be capable of perceiving such things. In effect, just as the eye was created to [see] colors, the ears to [hear] sounds, so the spirit of man was created to understand quantities..." (GW, vol. 13, p. 113). "Geometry... passed into the spirit of man at the same time as the image of God: it was not only through the eyes that it was received within" (GW, vol. 6, p. 223).

17. Wagner, "Théorie de l'image," p. 171.

18. Cf. Panofsky, *Idea*, ch. 5.

19. *Ibid.*, p. 84.

20. GW, vol. 1, p. 23. This theme is repeated frequently: GW, vol. 2, p. 19; vol. 7, p. 267; vol. 8, pp. 186ff.; etc.

21. Proclus, *Commentary on the First Book of Euclid's Elements*, trans. G. R. Morrow (Princeton, NJ: Princeton University Press, 1970), p. 67; cf. also p. 88.

22. According to Dietrich Mahnke, Kepler based his views of Nicholas of Cusa on the *De mathematica perfectione* and the *Complementum theologicum*, but had no knowledge of *De docta ignorantia* or *De quadratura circuli*. See Mahnke, *Unendliche Sphäre und Allmittelpunkt* (Halle: Niemeyer, 1937), pp. 130–32, 140–42.

23. The most interesting passages include GW, vol. 1, p. 33; vol. 2, p. 13; and vol. 7, p. 57.

24. J. Hübner, *Die Theologie Johannes Keplers* (Tübingen: Mohr, 1975), pp. 191-92.

25. Zorzi, *De harmonia mundi* (Paris: Berthelin, 1544), 1.3.2. St. Augustine had already written, "In the Father unity, in the Son, equality, in the Holy Spirit the union of unity and equality..." (*De doctrina christiana* 1.5). Cf. also Charles Bovelles, *De sapiente* 30, R. Klibansky (ed.), in E. Cassirer, *Individuum und Kosmos in der Philosophie der Renaissance*, pp. 353–54, 366. Bovelles adopts the same division as Nicholas of Cusa. Concerning all this, cf. Mahnke, *Unendliche Sphäre und Allmittelpunkt*, pp. 107, 142–43.

26. Cf. below, p. 200.

27. Gérard Simon, *Kepler astronome astrologue*, p. 163.

28. Cf. also GW, vol. 1, p. 307; vol. 3, p. 295; vol. 6, p. 55; vol. 15, pp. 249 and 258.

29. In the introduction to his German translation of *De harmonia mundi*, Max Caspar writes that the "roots" of Kepler's work are to be found in the Renaissance: "Their roots are located here, even if many beginnings prepare what is to come. The view of the universe that he designs is static, drawn from aesthetic principles. Beauty emanates from similarity." Cf. also Leo Spitzer, *L'Armonia del mondo* (Bologna: Il Mulino, 1963), p. 170.

30. Erwin Panofsky, *The Life and Art of Albrecht Dürer*, p. 254.

31. M. Steck, *Dürers Gestaltlehre* (Halle: Niemeyer, 1948), pp. 43–44.

32. For Dürer, who thereby associated himself with the skeptical current of the Renaissance, this question is beyond the reach of man: "I do not deny that someone can invent and express with the hand one image that is more beautiful than another, and demonstrate with good reasons why it is so, but it is impossible to reach supreme perfection, such that no one could do any better. For perfection is not within the forces of human understanding; only divine understanding has knowledge of it...." Cited from Dürer, *Quatre livres sur les proportions humaines*, trans. J. Meigret (Paris: Wechsel, 1557), fol. 102ᵛ.

CHAPTER VIII: THE COSMOGRAPHIC MYSTERY

1. This was contrary to the principle of plenitude as traditionally accepted: "In fact, the Physicists think that from the lower surface of lunary heaven to the tenth sphere there is no void between the celestial orbs, but each orb touches the next, and the bottom of the upper surface is completely united with the top of the lower surface" (GW, vol. 1, p. 47).

2. Based on verses traditionally attributed to Ptolemy. Cf. Friedrich Seck, "Johannes Kepler als Dichter," in *Internationales Kepler-Symposium Weil-der-Stadt* (Hildenheim: Gerstenberg, 1973), pp. 427–50.

3. For a useful synthesis, cf. E. J. Aiton, "Johannes Kepler and the *Mysterium cosmographicum*," *Sudhoffs Archiv* 61 (1977), pp. 173–94.

4. Barbaro, *Commentarius Vitruvio* (1556), p. 24. We should note that for Barbaro geometry is *madre del disegno*, just as it is for Kepler's God. Barbaro's commentary is the most important of the sixteenth-century commentaries on Vitruvius. Cf. V. Zoulov, "Vitruve et ses commentateurs," in *La Science au XVIᵉ siècle* (Paris: Hermann, 1960), pp. 67–90. For Vitruvius (*De architectura* 1.11.3) eurythmy is defined as "a gracious and pleasant experience in the arrangement of the members [*compositionibus membrorum*]." The "arrangement of the members" is explained in the *Mysterium cosmographicum* in terms of the regular polyhedrons.

5. Plato, *Timaeus*, p. 25 (35b–c): "And he began the division in this way.

First he took one portion (1) from the whole, and next a portion (2) double of this; the third (3) half as much again as the second, and three times the first; the fourth (4) double of the second; the fifth (9) three times the third; the sixth (8) eight times the first; and the seventh (27) twenty-seven times the first." The successive portions correspond therefore to 1, 2, 3, 4, 9, 27, or alternates between multiples of two and multiples of three. R. S. Westman draws the comparison between Plato and Kepler in "Kepler's Theory of Hypothesis and the Realist Dilemma," in *Internationales Kepler-Symposium Weil-der-Stadt*, pp. 45–47.

6. Cf. Rudolf Wittkower, *Architectural Principles in the Age of Humanism*, pp. 103–04.

7. Cf. Aristotle, *De caelo* 2.13.293a, and *Metaphysics* 986a.

8. Cf. Westman, "Kepler's Theory."

9. Cf. Plato, *Timaeus* 53c–54a. Concerning the prestige of the equilateral triangle, cf. also Luca Pacioli, *Divina proportione*, ed. C. Winterberg (Vienna: Graeser, 1896), p. 132. For Vitruvius, as Pacioli points out, the noblest part of the body, the head, was "formata in su la forma dela prima figura in le recte linee, cioe triangula equilatera...posta per fondamento e principio...dal nostro Euclide nel primo luogo del suo primo libro [formed atop the form of the primary figure in straight lines, that is an equilateral triangle...placed as foundation and beginning...by our Euclid at the outset of his first book]."

10. Dietrich Mahnke, *Unendliche Sphäre und Allmittelpunkt*, p. 136.

11. Cf. Cornelius Agrippa, *La Philosophie occulte ou la magie*, vol. 2, p. 27; or H. de Jong, *Michael Maier's "Atlanta fugiens"* (Utrecht: Schotanus & Jens, 1965), pp. 89–90 (concerning emblem 21). On Kepler and alchemy, cf. Karin Figala, "Kepler and Alchemy," *Vistas in Astronomy* 18 (1975), pp 457–69.

12. Claude Dubois, *Le Maniérisme* (Paris: Presses Universitaires de France, 1979), p. 83.

13. Cf. the notes in Paola Barocchi (ed.), *Scritti d'arte del Cinquecento*, vol. 2, pp. 1892–94.

14. On the ancient theories, cf. E. Sachs, *Die fünf platonischen Korper* (Berlin: Moellendorf, 1917). For the Renaissance, cf. S. K. Heninger, *Touches of Sweet*

*Harmony* (San Marino, CA: Huntington Library, 1974), pp. 108-15, and the bibliography, p. 140, n. 77.

15. According to *Harmonices mundi* (GW, vol. 6, pp. 80-81). In similar passages in both editions of the *Mysterium cosmographicum*, Kepler associates the icosahedron (and not the dodecahedron) with celestial matter.

16. GW, vol. 8, p. 59: "Even though this chapter is nothing more than an astronomical game and should not be considered a part of the work, but a digression, nonetheless the reader may compare it to Ptolemy's reasoning in both the *Tetrabiblos* and *Harmonics*: he will see that our [reasoning] is not inferior to his, and is perhaps better."

17. In Barocchi, *Scritti d'arte del Cinquecento*, vol. 2, p. 2115.

18. Foucault, *The Order of Things*, p. 50.

19. In his controversy with Fludd, Kepler nevertheless insisted on the completely hypothetical character of his interpretation: "I spoke in conjecture, for the terms 'by chance' and 'perhaps' are sprinkled throughout" (GW, vol. 6, p. 428).

20. In *Mysterium cosmographicum*, Kepler cites two editions of Euclid: one edited by Campanus (GW, vol. 1, p. 44) and the other by Foix de Candale (GW, vol. 1, p. 46). On the editions of ancient mathematicians used by Kepler, cf. especially Joseph E. Hofmann, "Über einige fachliche Beiträge Keplers zur Mathematik," in *Internationales Kepler-Symposlium Weil-der-Stadt*, pp. 261-84.

21. Preface to Billingsley's English translation of the *Elements*; the passage is cited by Heninger, *Touches of Sweet Harmony*, p. 109.

22. Proclus, *Commentary on the First Book of Euclid's Elements*, p. 57; cf. also *Commentaire sur le Timée*, ed. André Festugière (Paris: Vrin, 1966 *et seq.*), vol. 3, pp. 79-80, 104, 105, 111. Kepler's first mention of Proclus only dates from 1599 (GW, vol. 14, p. 63). In *Harmonices mundi*, he would recall this passage (GW, vol. 6, p. 18).

23. The question is broached in chs. 3-8 of *Mysterium cosmographicum* (GW, vol. 1, pp. 29-34).

24. Plato, *Republic* 527b.

25. Polyhedrons are broached in the *last* book of Dürer's *Quattuor institutionum geometricarum libri*. As for Jean Cousin's *Livre de perspective*, the title page clarifies the meaning of the frontispiece in the following manner: "This present figure demonstrates the five Regular Bodies of Geometry (which are deduced point by point at the end of this present book)...."

26. Cf. Gustav R. Hocke, *Die Welt als Labyrinth*, pp. 111-19.

27. Jacques Bousquet, *Le Maniérisme* (Neuchâtel: Ides & Calendes, 1964), pp. 104 and 106.

28. Vasari, *Lives of the Artists: A Selection*, trans. G. Ball (Harmondsworth: Penguin, 1965), p. 96.

29. André Chastel, "Marqueterie et perspective au XV^e Siècle," *Revue des Arts* (1953), pp. 141-54. Cf. also R. Clair, "Le Visible et l'imprévisible," in *Rétrospective Magritte...*(Charleroi: Graphing, 1978), p. 34: "in its alternation of black and white squares," the checkerboard perspective "is only the simplest variation of marquetry."

30. Galileo, *Scritti letterari*, ed. A. Chiari (Florence: Le Monnier, 1970), p. 493. I return to the Galileo-Kepler connection in Chapter 9.

31. Cf. D. P. Walker, "Kepler's Celestial Music," *Journal of the Warburg and Courtald Institute* 30 (1967), pp. 228-50 (especially 242-45).

32. Galileo, *The Assayer*, trans. S. Drake, in *The Controversy on the Comets of 1618: Galileo Galilei, Horatio Grassi, Mario Guiducci, Johann Kepler* (Philadelphia: University of Pennsylvania Press, 1960), p. 279. Altogether within the Mannerist spirit, Kepler conceived (but abandoned) the project of having a jeweler construct a model of the polyhedrons embedded in spheres. It would have been decorated with precious stones representing the planets, each of which could have been filled with a drink suitable for the "character" of the planet. A veritable curiosity, worthy of the *Wunderkammer* of the period. Cf. Ewa Chojeka, "Johannes Kepler und die Kunst," *Zeitschrift für Kunstgeschichte* 30 (1967), pp. 55-72.

33. Cf. especially Harold Coxeter, "Kepler and Mathematics," *Vistas in Astronomy* 18 (1973), pp. 661-70; and J. P. Phillips, "Kepler's Echinus," *Isis* 56 (1965), pp. 196-200. The metaphoric names are comparable to those found

in abundance during the same period, in Desargues for example; cf. Jean Mesnard, "Baroque, science et religion chez Pascal," *Baroque* 7 (1974), pp. 71-83 (especially 71-72).

34. Cf. John D. North, "Apian and Pacioli's Polyhedra," *Physics* (1965), p. 211-14.

35. Cf. especially Barbaro's *La Practica della perspettiva* (Venice: Burgominieri, 1568), pp. 111ff.

36. H. Tuzet, *Le Cosmos et l'imagination* (Paris: Corti, 1965), pp. 52-53.

CHAPTER IX: THINKING THE ELLIPSE

1. Eugenio d'Ors, *Du Baroque* (Paris: Gallimard, 1968), p. 10.

2. Severo Sarduy, *Barroco* (Paris: Éd. du Seuil, 1975), p. 58.

3. Otto Benesch, *The Art of the Renaissance in Northern Europe* (Cambridge, MA: Harvard University Press), 1945, pp. 137ff.; Hocke, pp. 133-39. Cf. also: Wylie Sypher, *Four Stages of Renaissance Style* (Garden City, NY: Doubleday, 1955), p. 134; James Ackerman, "Science and Visual Art," in S. Toulmin (ed.), *Seventeenth-Century Science and the Arts* (Princeton, NJ: Princeton University Press, 1961), especially pp. 72-73; and Arnold Hauser, *Mannerism* (New York: Knopf, 1964), vol. pp. 46-48. These references are only indicative, and do not pretend to be complete.

4. Erwin Panofsky, *Galileo as a Critic of the Arts* (The Hague: Nijhoff, 1954), p. 25.

5. *Ibid.*; Alexandre Koyré, "Attitude esthétique et pensée scientifique," in *Études d'histoire de la pensée scientifique*, pp. 275-88, Koyré's article appeared earlier in *Critique* (Sept.-Oct. 1955), pp. 835-47.

6. This is the case for those who like Hocke or Hugo Friedrich work in the tradition of Ernst Robert Curtius. See Curtius, *European Literature and the Latin Middle Ages* (New York: Pantheon Books, 1953); Hocke, *Die Welt als Labyrinth* and *Manierismus in der Literatur* (Hamburg: Rowoht, 1959); and Friedrich, *Epochen der italienischen Lyrik* (Frankfurt: Klosterman, 1964). Conversely, D'Ors and others posit the existence of baroque "periods," of which Mannerism of the sixteenth-seventeenth centuries becomes one instance (if it is not seen as

an avatar of a recurrent classicism, which is the case for Heinrich Wölfflin).

7. On these criteria, cf. my study, "La Correspondance dans les études interdisciplinaires," *Les Méthods du discours critique dans les études seiziémistes* (Paris: SEDES, 1987), pp. 39–52.

8. Cf. especially G. von Schwarzenberg, *Rudolf II* (Munich: Callwey, 1961).

9. GW, vol. 3, p. 36.

10. GW, vol. 3, pp. 7–10, about the war (presentation of *Astronomia nova* to Rudolf II); the rhetoric is primarily circumstantial. The image of navigation appears to be the most important; it appears numerous times, notably on pp. 36 (allusion to the Argonauts, Columbus, Magellan) and 270 (citation from the *Aeneid*). With war and navigation, Kepler took up the two great literary themes for the allegorical dramatization of action. Cf. Angus Fletcher, *Allegory: The Theory of a Symbolic Mode* (Ithaca, NY: Cornell University Press, 1964), Ch. 3. Cf. also Michel Serres, *Le Système de Leibnitz et ses modèles mathématiques* (Paris: Presses Universitaires de France, 1968), pp. 623ff. Serres discusses the transformation of the "odyssey," which in the modern era ceases to be circular – as is the case, at least emblematically, for *Astronomia nova*.

11. For a detailed analysis, cf. Gérard Simon, *Kepler, astronome astrologue*, chs. 6 and 7.

12. Expressions utilized by Jürgen Mittelstrass, *Die Rettung der Phänomene*, p. 218; and Simon, *Kepler, astronome astrologue*, p. 368.

13. GW, vol. 3, p. 366 ("we will demonstrate that no figure remains possible for the planetary orbit except the regular ellipse...").

14. Vincenzo Danti, *Trattato delle perfette proporzione*, in Paola Barocchi (ed.), *Scritti d'arte del Cinquecento*, vol. 2, p. 1769.

15. Borghini, *Il Riposo*, in Barocchi, *Scritti d'arte del Cinquecento*, vol. 2, p. 1839.

16. Curtius Wilson, "Kepler's Derivation of the Elliptical Path," *Isis* 59 (1969), pp. 13–14.

17. Such a conception still marked Copernicus' physics. Cf. Koyré, *The Astronomical Revolution*, p. 59.

18. *Il Riposo*, in Barocchi (ed.), *Scritti d'arte del Cinquecento*, vol. 2, p. 1839.

19. Lomazzo, *Trattato dell'arte*, p. 23 (1.3). Unlike Kepler, Lomazzo still thought in terms of Aristotelian physics. What is important from our point of view, on this celebrated page, is the value ascribed to dynamic relations: whatever the supporting theory, it illuminates a general form of receptivity.

20. John Shearman, *Mannerism* (Harmondsworth: Penguin Books, 1967), p. 83.

21. *Ibid.*, p. 86.

22. Benesch, pp. 141 and 160, especially concerning the *Martyrdom of Saint Maurice and the Theban Legion*.

23. Cf. Panofsky, *Idea* pp. 79ff. ("Dualism in Mannerist Thought"); and Sypher, *Four Stages of Renaissance Style*, pp. 162ff. ("Unresolved Tensions").

24. This is a traditional theme, as Georges Poulet recalls in *Les Métaphorphoses du cercle* (Paris: Plon, 1960), p. xx:

> For a long time, the Pythagoreans had insisted on the generative power of the point. For Plotinus, the center is the "father of the circle." For Scotus Erigena, it is the "universal initial point." Likewise, the Jewish Kabbala and Arab thought make the central point the focus from which the universe develops. All of the medieval philosophy of light is a long commentary on the spherical diffusion of any luminous point. *Omne agens multiplicat suam virtutem sphaerice* [Every agent multiplies its strength spherically]. This statement by Grosseteste applies to any activity, but primarily to God. God is a light that propagates, a power that increases and spreads."

25. Cf. Gerald Holton, *The Scientific Imagination*, p. 65.

26. Mittelstrass, "Wissenschafstheoretische Elemente der Keplerische Astronomie," in *Internationales Kepler-Symposium Weil-der-Stadt*, p. 26.

27. Hans Blumenberg, *Paradigmen zu einer Metaphorologie*, p. 177.

28. Kepler, *Paralipomènes à Vitellion*, trans. C. Chevalley (Paris: Vrin, 1980), p. 221. It is true that in this passage, Kepler establishes no hierarchical or symbolic distinction. But what appears here as a continuum is exploited symbolically elsewhere. Cf. the next citation.

29. Cf. Marjorie Hope Nicolson, *The Breaking of the Circle* (New York: Columbia University Press, 1960), p. 154; and H. Tuzet, *Le Cosmos et l'imagination*, p. 53.

30. Kepler, *Paralipomènes à Vitellion*, p. 221.

31. Blaise Pascal, *Traités des coniques*, in *Oeuvres complètes*, L. Lafuma (ed.) (Paris: Éd. du Seuil, 1963), p. 40.

32. Cf. E. Melandri, *La Linea e il circolo* (Bologna: Il Mulino, 1968), pp. 1062–63: "Moreover, if our geometry has as basic primary figures the line and the circle – construction by means of 'ruler and compass' is a requirement of ancient mathematics – a figure like the ellipse appears in and of itself irrational: in fact it is neither circle nor line. It can be produced only mechanically, not geometrically...by sectioning cones or using special instruments that are not within the scope of the compass." Kepler described how to construct an ellipse with the help of two pieces of string (*Paralipomènes à Vitellion*, p. 224); cf. the commentary by William Ivins, *Art and Geometry* (Cambridge, MA: Harvard University Press, 1946), ch. 7.

33. For Copernicus, the combination of the curve and straight line is possible with the participation of different motions, for example, for a terrestrial object that returns to its natural place while accompanying the earth in its global motion. Such motion is an exception, a return to order, rather than the order itself.

34. E. Dehennin, *Antithèse, oxymore, paradoxisme* (Paris: Didier, 1973), p. 36.

35. Leon Battista Alberti, *De pictura*, in *On Painting, On Sculpture*, trans. Cecil Grayson (London: Phaidon, 1972), pp. 83 and 85.

36. Hocke, *Manierismus*, p. 244. Cf. also Marcel Raymond, *La Poésie française et le Maniérisme* (Geneva: Droz, 1971), p. 35: "If we closely examine the idea that directed the practice of the Mannerist poets and justified their deviations in their own eyes, we find almost everywhere, either explicitly or implicitly, the idea of contradiction."

37. Robert Haas, "Keplers Weltharmonik in Vergangenheit, Gegenwart und Zukunft," *Sudhoffs Archiv* 57 (1973), pp. 41–70.

38. On the disappearance in the seventeenth century of such anthropomorphic representations, cf. Léon Brunschvicg, *L'Expérience humaine et la causalité physique* (Paris: Alcan, 1922), ch. 14.

39. Compared to the fundamental principles of astronomy, Kepler's innovations can appear more daring or shocking than Copernicus'. Cf. Mittelstrass, "Wissenschaftstheorische Elemente...." On the problematic relation for the Mannerists between the tangible and intelligible, cf. above, Chapter 6.

40. Simon Stevin, *De Hemelloop*, trans. C. Dikshoorn, in *The Principal Works of Simon Stevin*, vol. 3, *Astronomy and Navigation*, A. Pannekoek and E. Crone (eds.) (Amsterdam: Swets & Zeitlinger, 1961).

41. On Stevin, see Eduard J. Dijksterhuis, *Simon Stevin: Science in the Netherlands around 1600* (The Hague: Martinus Nijhoff, 1970). On Stevin and Copernicus in particular, see Alois Gerlo, "Copernicus en Simon Stevin," *Sartonia* 20 (1973), pp. 3-18 (especially the letter cited on p. 13). It is possible that Stevin hesitated before publishing his defense of Copernicus (Book 3 of *De Hemelloop*), because he confessed in Book 2, ch. 5 that he had put it aside for a time.

42. Stevin, *De Hemelloop*, in *The Principal Works of Simon Stevin*, vol. 3, p. 117 ("Summary of this Third Book").

43. Cf. Stevin, *De Hemelloop* 3.1.2-4.

44. *Ibid.* 3.1.5.

45. Cf. Panofsky, *Galileo as a Critic*, pp. 20ff.

46. *Ibid.*, p. 25.

47. Cf. Koyré's thesis on the Platonic character of the new science in general, and Galileo in particular: "Galilée et Platon," in *Études d'histoire*, pp. 166-95. This thesis has led to a long debate, which among other benefits has exposed the *differences* between Galileo and Plato; cf. especially Stillman Drake, "Galileo semantico," *Intersezioni* (1983), pp. 45-55.

48. Maurice Clavelin, *La Philosophie naturelle de Galilée* (Paris: A. Colin, 1968), p. 244.

49. Cf. above, Chapter 7.

50. Clavelin, *La Philosophie naturelle de Galilée*, p. 244.

51. J. L. Russell, "Kepler's Laws of Planetary Motion: 1609–1666," *British Journal for the History of Science* 2 (1964), p. 4.

52. "La Science et la désallégorisation du mouvement," in G. Kepes (ed.), *Nature et art du mouvement* (Brussells: La Connaissance, 1968), p. 26.

53. Pietro Redondi's *Galileo eretico* (Turin: Einaudi, 1983) offers a fine evocation of this milieu, and especially its ambiguities.

54. Mersenne, *Quaestiones in Genesim*, col. 58.

55. *Ibid.*, col. 88.

56. Cited and commented on by Ezio Raimondi, *Letteratura barocca* (Florence: Olschki, 1961), p. 21.

57. Mersenne, *De l'Utilité de l'harmonie*, in *Harmonie universelle* (Paris: Éd. du CNRS, 1963), vol. 3, pp. 21–22.

58. *Ibid.*, vol. 3, p. 16.

59. Book 1 ("On the voice"), *ibid.*, vol. 1, p. 32.

60. Cf. Giulio Carlo Argan, *L'Europe des capitales*, p. 70.

61. For the elaboratation of a parallel, cf. my article, "Tesauro vs. Port-Royal," *Baroque* 10 (1980), pp. 76–86.

62. Tesauro, *Il Cannocchiale aristotelico* (Venice: Milocho, 1682), p. 399. Cf. Lomazzo, cited above: a figure for what is spoken in *Trattato dell'arte*, the cone becomes a figure for the act of speaking in *Cannocchiale*.

63. *Ibid.*, pp. 164–65.

64. On such elliptical structures, cf. Sarduy, pp. 54, 73ff., 135ff.

65. Pascal, "Man's Disproportion" (Lafuma #199, Brunschvicg #72), *Pensées*, trans. W. F. Trotter (New York: E. P. Dutton, 1958), p 16. On Pascal's use of such procedures, cf. Serres, *Le Système de Leibnitz et ses modèles mathématiques* 3.2.

66. On the theme of perspective in Pascal, cf. Louis Marin, *Études sémiologiques* (Paris: Klincksieck, 1971), pp. 166ff.

67. Gérard Genette, *Figures* (Paris: Éd. du Seuil, 1966), p. 30.

68. *Ibid.*, p. 37.

69. Pascal, *Pensées*, p. 8 (Lafuma #559, Brunschvicg #27).

70. Raimondi, *Letteratura barocca*, pp. 24–32.

71. Tesauro, *Il Cannocchiale*, p. 66.

72. Descartes, *Oeuvres*, ed. C. Adam and P. Tannery (Paris: Léopold Cerf, 1897-1910), vol. 6, p. 392.

73. *Ibid.*

74. *Ibid.*, 2, p. 499.

75. Cited by Brunschvicg, *Les Étapes de la philosophie mathématique* (Paris: Vrin, 1972), p. 119.

76. Descartes, *Oeuvres*, vol. 6, p. 475.

77. Cf. the discussion on the analysis of conic sections in Book 1 of *Harmonices mundi* (GW, vol. 6, pp. 53ff.).

78. Descartes, *Oeuvres*, vol. 7, p. 139.

79. On the importance of this activity for Descartes, see J.-L. Nancy, *Ego sum* (Paris: Flammarion, 1979). The literary rhetoric of the Baroque also placed a premium on activity. Cf., G. Conti, *La Metafora barocca* (Milan: Mursia, 1972), pp. 69-86; the final goal ("il fare ingegnoso") is of course different.

80. This chapter is a revised and expanded version of a contribution to *Manierismo e Letteratura*, ed. Daniela Dalla Valle (Florence: Meynier, 1987), pp. 157-76.

CHAPTER X: THE MUSICAL METAPHOR

1. Robert Haas, "Keplers Weltharmonik in Vergangenheit, Gegenwart und Zukunft," pp. 41-70.

2. The bibliography is immense. Two recent studies: Dorothy Koenigsberger, *Renaissance Man and Creative Thinking: A History of Concepts of Harmony* (Hassocks: Harvester Press, 1980); and H. Schavernach, *Die Harmonie der Sphären* (Munich: Albert Freiburg, 1981).

3. Zarlino, *Istitutioni armoniche* (Venice, Franceschi, 1573), and *Sopplimenti musicali* (Venice: Franceschi, 1588). Before Zarlino, Fogliniani's *Musica theorica* (1529) was already moving in the same direction. Rudolf Wittkower has shown that ratios like 4:5 or 5:6 also appear in Palladio's architecture. See Wittkower, *Architectural Principles in the Age of Humanism* (New York: Norton, 1971), pp. 132ff. Kepler refers to Zarlino in *Harmonices mundi* (GW, vol. 6, p. 139).

4. Vincenzo Galilei is the modern authority in musical matters to whom Kepler refers the most often. Cf. D. P. Walker, "Kepler's Celestial Music," *Journal of the Warburg and Courtauld Institute* 30 (1967), p. 233.

5. On the distinction between somatopoetic figures and cosmopoetic figures, cf. above, p. 180.

6. Besides knowability, Kepler retained disposition for congruence as a criterion for classification. Congruence can be obtained in either two or three dimensions; it is not related to the harmony of celestial motion but to astrological influence.

7. Cf. Gérard Simon, *Kepler astronome astrologue*, p. 150.

8. Kepler found himself confronting problems of inclusion (for the pentagon and octagon) and exclusion (for the pentadecagon). Cf. Walker ("Kepler's Celestial Music"), who stresses the importance of passages on the pentagon, divine proportion, and the sexual symbolism of geometric figures.

9. Michael Riffaterre, *La Production du texte* (Paris: Éd. du Seuil, 1979), p. 223.

10. For a detailed study, cf. H. Atteln, *Das Verhältnis Musik-Mathematik bei J. Kepler*, thesis (Erlangen-Nuremberg: Friedrich-Alexander Universität, 1970); and Michael Dickreiter, *Der Musiktheoretiker J. Kepler* (Berne: Francke, 1973).

11. Cornelius Agrippa, *La Philosophie occulte ou la magie*, vol. 2, p. 26.

12. In relation to the moon as well as Venus.

13. In relation to Saturn.

14. The third law affirms that the ratio of the squares of the periods of revolution to the cubes of the distances is constant: $T^2/R^3 = c$. On the significance of this law for Kepler, cf. especially V. Bialas, "Die Bedeutung des dritten Planetengesetzes für das Werk von J. Kepler," *Philosophia naturalis* (1971), pp. 42–55; and Haas, "Marginalien zum 3. Keplerischen Gesetz," in E. Preuss (ed.), *Kepler-Festschrift* (Regensburg: Naturwissenschaftlicher Verein, 1971), pp. 159–65.

15. Cf. GW, vol. 6, p. 358, or for a simplified table, Koyré, *The Astronomical Revolution*, p. 340.

16. Cf. Koyré, *The Astronomical Revolution*, p. 341.

17. GW, vol. 7, p. 282. All this is combined with the choice of the earth as the central vantage point for viewing the universe; cf. above, the second section of this chapter, "The Song of the Planets."

18. Koyré, *The Astronomical Revolution*, p. 355. For comparable tables, specifically alchemical, cf. Robert Lenoble, *Mersenne ou la naissance du mécanisme* (Paris: Vrin, 1943), pp. 134–35.

19. Plato, *Timaeus*, p. 21 (32c).

20. On Daedalus as a Mannerist archetype, cf. Gustave Hocke, *Manierismus in der Literatur*, pp. 204–14.

21. Cf. Peter J. Ammann, "The Musical Theory and Philosophy of Robert Fludd," *Journal of the Warburg and Courtauld Institute* 30 (1967), pp. 198–227.

22. Yates, *Giordano Bruno*, p. 406.

23. Kepler entered into the debate in an appendix to *De harmonice mundi* (1619). Fludd replied in *Veritatis proscenium* (1621). Kepler countered with *Apologia* (1622), and Fludd published his *Monochordum mundi* the same year.

24. Ammann, "The Musical Theory and Philosophy of Robert Fludd," pp. 223–24.

25. Cf. Schavernach, *Die Harmonie der Sphären*, p. 150.

26. Cf. Athanasius Kircher, *Musurgia universalis* (Rome: Corbelletti, 1650) vol. 2, p. 377: Kepler "constructed another harmony, but so obscure and so shrouded in the mystical veil of words that it is difficult to understand what he thinks...."

27. Cf. above, p. 167.

28. Cf. Ammann, "The Musical Theory and Philosophy of Robert Fludd," pp. 216–18; and Lenoble, *Mersenne ou la naissance du mécanisme*, p. 107.

29. Fludd, *Veritatis proscenium*, appendix to the 2nd ed. of *Utriusque cosmi... metaphysica, physica atque technica historia*, vol. 2 (Frankfurt, 1621), p. 10: "Quite to the contrary, the science of all the sounds of music, although these arise from intervals, more concerns physics because of their occult nature, than mathematics, if we examine appropriately the internal principle of this same music."

30. *Ibid.*, p. 15.

31. *Ibid.*

32. *Ibid.*, p. 10.

33. *Ibid.*, p. 15.

34. *Ibid.*, p. 29: "We say therefore that our contemplation of the pyramid in mathematical terms is more formal than comprehensible, since it reveals the hidden progression of form in matter towards purification, rarefaction, and progress towards maturity and perfection from imperfection, and from the thick and dense towards the subtle...."

35. Cf. Giulio Carlo Argan, *L'Europe des capitales*, p. 31: "At the philosophical level, the Baroque doubtless appears as a reaction against an idealist philosophy and the Neoplatonism of the Mannerists, as a return to the philosophy of experience.... The imagination had reached the level of mental activity: a mode of thinking with images rather than with logic."

36. Cf. the end of the work: "Epilogus sive Metaphysica lucis et umbrae" (pp. 796-807), and "Aphaera mystica sive Tropologia lucis et umbrae," where the "tropological rules" are explained "according to which the soul, emerging from shadow towards the light, unites with the perfectly eternal light from which it flows" (p. 808). We find the combined influence of Fludd and Kircher in the fine cosmological poem of Juana Inés de la Cruz, *Primero sueño* (1692); cf. especially Octavio Paz, *Sor Juana Inés de la Cruz* (Barcelona: Seix Barral, 1982), pp. 469-507. Fludd's two pyramids (of spirit and matter) are presented most notably in his *Oedipus Aegyptiacus* (Rome, 1652-1654), vol. 2, p. 115; and in Kircher's *Musurgia universalis* (vol. 2, p. 393); as well as in the *Sueño* of the Mexican nun.

37. On *Ars magna*, cf. Claude Chevalley, "*L'Ars magna lucis et umbrae* d'Athanase Kircher: 1646 and 1671," *Baroque* 12 (1987), pp. 95-110. On Kircher's work in general, cf. Jocelyn Godwin, *Athanasius Kircher* (London: Thames & Hudson, 1979), with bibliography.

38. On the importance of synesthesia for Kircher, cf. René Wellek, "Renaissance und Barocksynästhesie," *Deutsche Vierteljahrschrift für Literaturgeschichte* 9 (1930), p. 559.

39. Kircher, *Musurgia* 10.2, vol. 2, p. 381.

40. *Almagestum novum* 9.5.7, vol. 2, p. 531. Coming from a different direction, Spinoza's criticism in *The Ethics* is all the more severe: "there are men lunatic enough to believe, that even God himself takes pleasure in harmony; and philosophers are not lacking who have persuaded themselves, that the motion of the heavenly bodies gives rise to harmony...men have lost their reason over harmony, going so far as to believe that God had also been similarly enraptured! There have even been philosophers who believe that celestial motions comprise a harmony...." *The Chief Works of Benedict Spinoza*, trans. R. H. M. Elwes (New York: Dover, 1955), vol. 2, p. 80.

41. *Book One of the Instruments*, prop. xi, in *Harmonie universelle*, vol. 3. Mersenne does not exclude, however, the musical composition of the universe; he merely declares that it is inaccessible to man: *Book Two of the Motion of Bodies*, prop. vi (*Harmonie universelle*, vol. 1).

42. Cf. above, pp. 224-25.

43. Argan, *L'Europe des capitales*, p. 21.

44. Tesauro, *Il Cannocchiale aristotelico*, p. 51.


CHAPTER XI: THE LUNAR *Dream*

1. *Joh. Keppleri mathematici olim imperatorii Somnium seu opus posthumum de astronomia lunari. Divulgatum a M. Ludovico Kepplero filio medicinae candidato. Impressum partim Sagani Silesiorum, absolutum Franconforti, sumptibus haeredum authoris. Anno MDCXXXIV* [*The Dream or posthumous work on lunar astronomy of Johannes Kepler late imperial mathematician. Published by his son Ludwig Kepler, M.A., candidate for the doctorate in medicine. Printed in part at Zagan in Silesia, completed at Frankfurt, at the expense of the author's heirs. 1634*]. With an *appendix geographica, seu mavis, selenegraphica...* [*a geographical, or if you prefer, selenegraphical appendix*], and an annotated Latin translation of Plutarch's *Libellus de facie quae in orbe Lunae apparet* [*The Face in the Moon*]. The work is included in Frisch's edition of the *Opera omnia* (vol. 8, pp. 27ff). It has not yet appeared in the *Gesammelte Werke*. A facsimile of the original edition has been published by M. Liszt and W. Gerlach (Osnabrück, 1969).

A German translation exists by L. Guenther, and an English translation by P. F. Kirkwood, accompanied by a study by J. Lear. See Guenther (trans.), *Keplers Traum von Mond* (Leipzig: Teubner, 1898); and Kirkwood (trans.), *Kepler's Dream* (Berkeley: University of California Press, 1965). I cite from Edward Rosen's English translation: *Kepler's Somnium* (Madison: University of Wisconsin Press, 1967).

2. On this question and for information of a historical nature on the *Somnium*, cf. Rosen's edition and three works by Marjorie Hope Nicolson: *A World in the Moon* (Northampton, MA: Smith College, 1936); *Voyages to the Moon* (New York: MacMillan, 1948); and *Science and Imagination* (Ithaca, NY: Cornell University Press, 1956). On the relations between Kepler and Donne, cf. also Wilbur Applebaum, "Donne's Meeting with Kepler: A Previously Unknown Episode," *Philological Quarterly* 50 (1970), pp. 132–34.

3. On lunar literature, cf. the works cited above by Nicolson. For a recent and substantial bibliography, cf. Darko Suvin, *Pour une poétique de la science-fiction* (Montreal: Presses de l'Université du Québec, 1977). On Kepler's influence, see the synthesis of M. Ducos in the introduction to her French translation of the *Dream: Le Songe, ou Astronomie lunaire* (Nancy: Presses Universitaires de Nancy, 1984), pp. 14–18.

4. Cf. Shearman, *Mannerism*, pp. 92–96. La Ceppède provides another example from this period of highly developed self-commentary in his two books of *Théorèmes* (1613, 1621); each sonnet is accompanied by a gloss of several pages.

5. Jacques Maritain, *The Dream of Descartes, Together with some other Essays* (New York: Philosophical Library, 1944); and Georges Poulet, *Studies in Human Time* (Baltimore: Johns Hopkins Press, 1956), ch. 2. On the narrative of Descartes' dream in relation to the tradition of oneiromancy, cf. J.-M. Wagner, "Esquisse du cadre divinatoire des songes de Descartes," *Baroque* 6 (1973), pp. 81–95.

6. Adrien Baillet, *Vie de Monsieur des Cartes* (Paris: Horthemels, 1691). The dream narrative is on pp. 81–86 (2.1).

7. On Artemidorus, see OO, vol. 1, p. 393. On Cicero's *De divinatione*, see GW, vol. 4, p. 251; vol. 16, p. 261. Notes on Ptolemy's *Harmony* contain numer-

ous references to Macrobius (OO, vol. 5, pp. 340, 348, 362, 382, 397, 411). On Peucer, see GW, vol. 16, p. 377. These references are purely indicative, not exhaustive.

8. Ambrosius Macrobius, *Commentary on the Dream of Scipio* (New York: Columbia University Press, 1952), pp. 87–90. In the copious literature on oneiromancy, we cite the following for its discussion of taxonomy: A. H. M. Kessels, "Ancient Systems of Dream-Classification," *Mnemosyne* 4.22 (1969), pp. 389–424.

9. Cornelius Agrippa, *La Philosophie occulte ou la magie*, vol. 1, p. 167 (1.59). For Kepler on the subject of Agrippa, see GW, vol. 16, p. 384.

10. Peucer, *De praecipius generibus divinationum*...(Wittenberg: Schuvertel, 1560), p. 234. For a historical perspective centered on the Renaissance, cf., among others, Luciano Bonuzzi, "Qualche considerazioni su sonno e sogno dal tramonto della cultura classica all'età del manierismo," *Acta Medicae Historae Patavina* 20 (1973-1974), pp. 57-104; and Alice Browne, "Girolamo Cardano's *Somnium Synesiorum Libri III*," *Bibliothèque d'Humanisme et Renaissance* 41 (1979), pp. 123-35.

11. Macrobius, *Commentary*, p. 90.

12. *Ibid.*, pp. 83-87.

13. On this point, cf. Hans Blumenberg, *Genesis of the Copernican World* (Cambridge, MA: MIT Press, 1987), pp. 447-48.

14. "For Levania seems to its inhabitants to remain just as motionless among the moving stars as does our earth to us humans" (KS, p. 17).

15. Cited by Marcel Raymond, *La Poésie française et le Maniérisme*, p. 24.

16. Rosen, "An Alleged Inconsistency in Kepler's *Somnium*," *The Classical Outlook* 45.3 (Nov. 1967), pp. 28-29.

17. Cf. Plutarch's *De facie* (in OO, vol. 8, p. 102) and *De genio* (591a-c).

18. Georges Méautis, "Le Mythe de Timarque," *Revue des études anciennes* 52 (1950), p. 206. Méautis also emphasizes the Pythagorean origin of the motif.

19. Cf. Frye, *Anatomy of Criticism*, p. 161; and O. B. Hardison's remark concerning Milton and heliocentric theory in Gingerich (ed.), *The Nature of Scientific Discovery*, pp. 474-75.

20. Cf., for example, Peucer, *De praecipius generibus divinationum*, pp. 234–65, for a very detailed description of how images are formed in dreams.

21. Cicero, *De divinatione* 2.

22. GW, vol. 14, p. 432: "At somnia figuris et tropis constant oratoriis."

23. Aristotle, *On Prophecy in Sleep*, in *On the Soul, Parva Naturalis, On Breath*, trans. W. S. Hett (Cambridge, MA: Harvard University Press, 1964), p. 385.

24. KS, pp. 35–36 (n.3). As Rosen notes in his translation, *Fiolx* is probably a misreading of *fjord*.

25. Aristotle, *On Prophecy in Sleep*, p. 377.

26. KS, pp. 62 and 63.

27. Cf., for example, the interpretation Descartes gave to his dreams according to Baillet: the dictionary = a gathering of all the sciences; the *Corpus poetarum* = the uniting of philosophy and wisdom; the melon = love of solitude.

28. KS, p. 59 (n.38). See also p. 50, n. 35 concerning why "the most important" daemons are nine in number.

29. It is true that some writings on dreams prohibit such a double interpretation. This is the case for Cardano; cf. Browne, "Girolamo Cardano's *Somnium Synesiorum Libri III.*"

30. Cf. Gilbert Durand, *Les Structures anthropologiques de l'imaginaire* (Paris: Bordas, 1969), pp. 100–01, 255–56, 318–19.

31. Bachelard, *Psychoanalysis of Fire*, p. 11.

32. *Ibid.*, p. 12.

33. Durand, *Les Structures anthropologiques de l'imaginaire*, p. 327.

34. Cf. for example Plato's *Symposium* in *The Dialogues of Plato*, trans. B. Jowett (New York: Random House, 1937), vol. 1, p. 328 (202e): " 'He is a great spirit (*daimon*), and like all spirits he is intermediate between the divine and the mortal.' 'And what,' I said, 'is his power?' 'He interprets,' she replied, 'between gods and men, conveying and taking across to the gods the prayers and sacrifices of men, and to men the commands and replies of the gods….' "

35. In addition to the works cited by Nicolson, cf. Donald H. Menzel,

"Kepler's Place in Science Fiction," *Vistas in Astronomy* 18 (1975), pp. 896–904; and Gale E. Christianson, "Kepler's *Somnium*: Science Fiction and the Renaissance Scientist," *Science Fiction Studies* (1977), pp. 79–80.

36. According to Suvin (*Pour une poétique de la science-fiction*), extrapolation and analogy, knowledge ahead of its time and a utopian connection are the dominant traits of science fiction.

37. A dissertation subject proposed in 1593 at the University of Tübingen concerned the perception of celestial phenomena from the moon. Cf. OO, vol. 8, p. 677.

38. Galileo, *Dialogue on the Great World Systems*, p. 74.

39. Cyrano de Bergerac, *Histoire comique des éstats et empires de la lune et du soleil*, ed. C. Mettra and J. Suyeux (Paris: Pauvert, 1962), p. 19. Cf. C. Barbe, "Cyrano: La Mise l'envers du vieil univers d'Aristote," *Baroque* 7 (1974), pp. 49–70. On Kepler's influence on Cyrano, cf. Nicolson, *Voyages to the Moon*; and Erica Harth, *Cyrano de Bergerac and the Polemics of Modernity* (New York: Columbia University Press, 1970).

40. Graciàn, *El Criticón*, in A. Del Huyo (ed.), *Obras completas* (Madrid, Aguilar, 1960), p. 593 (1.8). On the vogue of this theme during the period, cf. Jean Lafond and Augustin Redondo (eds.), *L'Image du monde renversé et ses représentations littéraires de la fin du XVIᵉ siècle au milieu du XVIIᵉ* (Paris: Vrin, 1979).

41. Louis Marin, *Utopiques, jeux d'espace* (Paris: Éd. de Minuit, 1973), p. 110.

42. On Cigoli's drawing, cf. Erwin Panofsky, *Galileo as a Critic of the Arts*; and Severo Sarduy, *Barroco*.

43. Marin, *Utopiques, jeux d'espace*, p. 76.

44. Cited by Jurgis Baltrusaitis, *Anamorphoses ou magie artificielle des effets merveilleux* (Paris: Perrin, 1969), p. 159.

45. André Festugière, *La Révélation d'Hermès Trismégiste*, vol. 1, *L'Astrologie et les sciences occultes* (Paris: Gabalda, 1950), pp. 309ff.

46. Galileo, *Dialogue on the Great World Systems*, p. 469.

47. Cf. also Timothy J. Reiss, *The Discourse of Modernism*, ch. 4 ("Kepler,

His *Dream* and the Analysis and Pattern of Thought").

48. An earlier version of this chapter appeared in *Bibliothèque d'Humanisme et Renaissance* 42 (1980), pp. 329–47.

## CONCLUSION

1. Kuhn, *The Copernican Revolution*, p. 170.
2. Simon, *Kepler astronome astrologue*, pp. 459–60.

# Index

ABDUCTION, 7–8 and n.1, 13–14, 16, 29.

Accolti, Pietro (?–1627), 103n.66.

Ackermann, James, 204n.3.

Africa, Thomas, 62n.25.

Agrippa von Nettesheim, Heinrich Cornelius (1486–1534), 27, 74n.3, 188n.9, 189, 236, 246, 257 and n.9; *On the Uncertainty and Vanity of the Arts and Sciences*, 51.

Aiton, E. J., 185n.3.

Alberti, Leon Battista (1404–1472), 69, 70, 72, 94, 97, 103, 116; *De pictura*, 68n.43, 100–01, 103n.69, 215; *De re aedificatoria*, 94 and n.42, 100–01.

Algebra, 229–30; *see also* Astronomy, Geometry, Mathematics.

Allegory, 236, 257ff., 267ff.

Allers, Rudolf, 74n.2.

Amico, Giovanni Battista (?–1538), 76n.8.

Ammann, Peter, 247, 248n.28.

Ammirato, Scipio (1531–1601), 160.

Anagogy, 21ff., 25, 35, 54ff., 152, 158–59, 161, 216, 270–71, 282.

Analogy, 57, 114 and n.21, 146–47, 189ff., 251; *analogia*, 95; *analogon*, 146.

Anamorphosis, 101–03; literary, 278–79.

*anima*, 173–74.

Anomaly, 9.

Anthropocentrism, 57–58; *see also* Homocentrism, Theocentrism.

Anthropomorphism, 218n.38.

Antiquity, 59, 62ff., 105, 120, 129, 151; *see classical authors by name.*

Antithesis, 210, 215, 227ff.

Apollo, 132.

Apollonian ideal, 103, 209, 247–48.

Apollonius of Pergamum (*ca.* 260–*ca.* 180 B.C.), 76.

Archilochus, 48.

Architecture, ecclesiastical, 27, 137–39; elliptical, 226; radial, 138–39.

Argan, Giulio Carlo, 103nn.68, 69, 113, 117, 160n.31, 226n.70, 250n.35, 251n.43.

Aristarchus of Samos (*ca.* 310–*ca.* 230 B.C.), 60n.14, 193.

Aristotelianism, 47, 112, 113, 115 and n.35, 152, 163 and n.1, 172.

Aristotle, 14, 40, 44n.28, 47n.39, 49, 67, 76, 92–93 and n.35, 106–07, 109–10, 113n.18, 131, 133–36, 191–93, 199, 232, 265, 267; *De caelo*, 40n.16, 74–75, 92–93, 110n.5, 131n.15, 136n.23, 186n.7, 191; *Poetics*, 13–14.

Artemidorus of Ephesus (2nd C), 256n.7; *Key to Dreams*, 256. *artifex*, 57.
Asclepius, 64–65.
*Asclepius*, 128n.5, 136n.25.
Astronomy, 23, 37–38, 47, 54, 55–58, 63–67, 73–74, 125, 146, 151, 159, 173, 201–02, 209, 219n.39; geometric (Greek), 12; mathematical, 38–40, 50–51, 56–67, 66–67, 110–11, 125, 155, 173, 283; physical, 110–11.
Atteln, H., 236n.10.
St. Augustine (354–430), 47, 171, 179n.25; *Contra academicos*, 47.
Averroës, Ibn Rusd (1126–1198), 40 and n.16, 47.

BACHELARD, GASTON, 11 and n.17, 31, 269.
Baillet, Adrien (1649–1706), *Vie de Descartes*, 255–56, 268n.27; 257.
Baldus, Bernadino (1553–1617), *Vitruvian Lexicon*, 94.
Baltrusaitis, Jurgis, 102n.64, n.65, 279n.24.
Barbaro, Daniel (1513–1570), 200; *Commentarius Vitruvio*, 186 and n.4; *Pratica della perspettiva* (1569), 181, 200n.35.
Barbe, C., 273n.39.
Barocchi, Paolo, 145n.44, 189n.13, 191n.17.
Baroque, 203–04, 224, 226–28, 230n.79, 247, 250 and n.35, 251–52.
Barthes, Roland, 30n.69; "writing degree zero," 93.
Baudelaire, Charles, 14.
Beaune, Florimond de (1601–1652), 229.
Bellenger, Y., 95n.48.
Benesch, Otto, 203, 210–11 and n.22.
Benveniste, Émile, 68.
Bernegger, Matthias (1582–1640), 274.

Bernhardt, Jean, 93n.37, 128.
Bérulle, Pierre de (1575–1629), 141–45.
Bible, 23, 49, 63, 65, 130, 139, 152, 155; *Ecclesiastes* (3:11), 59; *Genesis*, 62; *John* (12:45, 14:6), 179; *Psalms* (93:1), 41–42.
Bilinski, B., 62n.5.
Billingsley, Henry (*ca.* 1535–1606), 194.
Birkenmaier, A., 79n.13.
Black, Max, 28.
Blumenberg, Hans, 28–29, 40, 57n.5, n.7, 58n.10, 160n.30, 212 and n.27, 259n.27.
Boccaccio (1313–1375), 260.
Boethius (*ca.* 480–524), 232; *De placitis philosophorum*, 189.
Bohr, Niels, 8–9.
Bonuzzi, Luciano, 257n.10.
Borghini, Raphael (1541–1588), 207–08, 218.
Borromeo, Carolo (1538–1584), 145.
Bousquet, Jacques, 196n.27.
Bovelles, Charles de (1470–1553), 57–58, 179n.25, 188.
Bracelli, Giovanni Battista (1584–1609), 196.
Brahe, Tycho (1546–1601), 59, 93–94, 108, 123 and n.47, 129 and n.8, 154–56 and n.22, 158–59, 160, 161, 206, 216, 231, 241; in Kepler's *Dream*, 263, 265, 270.
Brémond, Henri, 143ff.
Brennger, Johann Georg (*fl.* 1594–1629), 163n.1.
Browne, Alice, 257n.10, 269n.29.
Brunelleschi, Filippo (1379–1446), 103–117; Santa Maria degli Angeli (Florence), 138.
Bruno, Giordano (1548–1600), 59n.11, 154–55, 160, 161; *De immenso, De infinito*, 155 and n.18; *La Cena de la ceneri*, 155.
Brunschvicg, Léon, 19, 218n.78,

229n.75.
Budé, Guillaume (1467-1540), 65n.34.
Buridan, Jean (1300-1358), 44–46 and
  n.30, 55, 61, 111 and n.9.
Burke, Kenneth, 22n.49, 23, 39n.13.
Busson, Henri, 111n.9, 153n.7.
Butterfield, H., 43.

CABALA, 165, 167, 248; *see also* Mysti-
  cism, Play.
Callippus, 76.
Calvin, John (1509-1564), 159n.29.
Campanella, Tomaso (1568-1639), 161;
  *City of the Sun*, 254, 274.
Campanus de Novare (2nd half of
  13th C), 194n.20.
Cardano, Jerome (1501-1576),
  269n.29.
Carneades, 48.
Caspar, Max, 181n.29.
Cassirer, Ernst, 57n.9, 179n.25.
Cauchy, Augustin-Léon, 200.
Center(s), 57–58, 130ff.; absolute,
  117ff.; and periphery, 132ff., 145ff.;
  and ellipse, 203ff.; *see also* Foci;
  multiplicity of, 91, 109ff., 115–17,
  131–32; of circle, 57, 118ff.; of
  sphere, 177–79; singular, 130ff.,
  137ff., 144, 145–47, 203, 211–12,
  221; symbolism, 131ff.; *see also*
  Equant, Geo- and Heliocentrism.
*certa ratio*, 68, 103; *see also* Ratios.
Cervantes, Miguel de (1547-1616), *Don
  Quixote*, 261.
Chastel, André, 115n.27, 198n.29.
Chevalley, Claude, 213n.28.
Chojeka, Ewa, 199n.32, 250n.37.
Christianson, Gale, 270n.35.
Chrysippus (*ca.* 280-206 B.C.), 96.
Church Fathers, 57; *see patristic authors
  by name*.
Cicero, Marcus Tullius (*ca.* 106-143
  B.C.), 59, 94; *De divinatione*, 256,
  265; *De natura deorum*, 64; *Repub-
  lic (Scipio's Dream)*, 253, 257–58.

Cigoli, Ludovico (1559-1613), 275
  and n.42.
Cipher, 103, 249.
Circle, 57–58, 76ff., 105ff., 114–15,
  118ff., 159, 176, 184, 205ff., 214,
  222ff., 233–34; *see also* Center,
  Motion, Space, Sun, Geo- and
  Heliocentrism.
Clagett, M., 61n.21, 92n.32, 111n.9.
Clair, R., 198n.29.
Clavelin, Maurice, 110n.7, 222,
  223n.50.
Clavius, Christophorus (1537-1612),
  59 and n.12.
Clichtove, Josse (2nd half of 15th
  C-1543), 63n.26.
Clitomachus, 48.
Code, 20, 67.
Codification, 20–21; *see also* Hyper-
  codification, Hypocodification.
Cohen, I. Bernard, 62n.24.
Columbus, Christopher (1451-1506),
  205n.10.
Commensurability, 84–89, 97, 114,
  118, 185; *see also certa ratio*, Incom-
  mensurability, Proportion.
*commensuratio*, 94.
Commonplaces, 16, 24.
*complicatio*, 116.
*concetti*, 198ff.
Cone, 210, 214, 227 and n.62.
Conic sections, 212, 214, 225,
  229n.77, 250.
Contextualism, 18, 25, 209–11, 220,
  284.
Continuity, 8ff.
*contrapposto*, 211, 215.
Cosmopoetic, 180–81, 182, 233.
Council of Trent, 145.
Counter Reformation, 145.
Cousin, Jean (*ca.* 1490-1561), 196
  and n.25.
Coxeter, Harold, 200n.33.
Crevier, Jean Baptiste, 16.
Crombie, Alistair, 43, 44n.26.

357

Culler, Jonathon, 25n.55.
Curtius, E. R., 204n.6.
Curve and line, 175ff., 180, 188, 195, 206, 213; *see also* Ellipse, Hyperbola, Parabola.
Cyrano de Bergerac, Savinien (1619–1655), *Estats et empires de la lune* (1657), 254, 273.

DAGENS, JEAN, 57n.5, 144 and n.40.
Dambska, I., 37n.4, 38n.11.
Damisch, H., 115n.27, 117n.35.
Dante Alighieri (1265–1321), 23n.51, 132n.16.
Danti, Vicenzo (1530–1576), 207, 219.
De Bruyne, Edgar, 96n.53.
De Coster, Michel, 28.
de Jong, H., 188n.11.
de Vocht, Henri, 65n.33.
*declamatio*, 50–52 and n.51.
*decorum*, 159ff.
Deferent, 76–77, 79, 83n.19, 84–85, 119; eccentric, 76, 81, 103, 119, 125.
Dehennin, E., 215n.34.
Delcourt, Marie, 50n.44.
Delfosse, Pascale, 12n.22.
Democritus, 48.
Desargues, Girard (1591–1661), 200n.33.
Descartes, René (1596–1650), 160n.30, 257; *Geometry*, 229–30.
Dickreiter, Michael, 236n.10.
Dicks, D. R., 93n.35.
*Didaskalikos*, 60 and n.16.
Discourse, 12–13, 22, 47n.36, 67, 69–70, 93.
*disegno*, 171–72 and n.10, 183, 186n.4, 191, 285.
Dobrzycki, Jerzy, 152n.1, 155n.19.
Dodoens, Rembert (1517–1585), 154; *Cosmographica isagoge*, 154 and n.12; *De sphaera*, 154 n.13, n.14.
Doland, E., 39n.12.
Donatello, Bonato di Berto di Bardo

(1383–1466), 197.
Donne, John (1573–1631), 253n.2; *Conclavium ignatii*, 253.
Drake, Stillman, 222n.47.
Dreams, five types (Macrobius), 256–57.
Dreyer, J. L. E., 94n.38.
Du Bartas, Guillaum de Saluste (1544–1590), 95n.48, 107–08.
Dubois, Claude, 188n.12.
Ducos, M., 254n.3.
Duhem, Pierre, 37n.4, 39n.12, 41, 43, 44n.26, 61n.21, 76n.10, 111.
Dumont, Jean Paul, 49n.43.
Dunbar, Helen, 129n.9.
Durand, Gilbert, 269n.30, 270.
Dürer, Albrecht (1471–1528), 93–94 and n.39, 96–97, 100 and n.57, 114, 156–57, 181–82 and n.32, 188, 196 and n.25.
Dynamism, 209–11.

ECKHART, JOHANNES (1260–1327), 46.
Eco, Umberto, 14, 20n.46.
*eidos*, 114, 172, 222.
Eisenstein, Elizabeth, 63n.31.
Elements, 11n.7, 106–07, 112–13, 115 n.28, n.29; 147, 189–90, 192–93, 197, 242.
Eliade, Mircea, 63n.30, 133.
Eliot, T. S., 28.
Ellipse, 199, 206, 208ff., 216–17, 220, 222ff., 284; hybrid of curve and line, 213ff.; as dynamic trace, 209f; *see also* Curve, Foci, Hyperbola.
Emmius, Ubbo (1574–1625), 220.
Empiricism, 11.
Empyrean heaven, 137.
Epicycle, 38, 76–80 and n.12, n.15; 83–89, 103, 114, 123, 147, 172, 283.
Equant, 76, 106, 118ff., 146–47, 155, 205, 283.
Equicola, Mario (1460–1539), 94 and n.43.

Erasmus, Didier (1467–1539), 46, 63, 65–67 and n.34, 94, 139, 165, 258; *Annotations* to New Testament, 66; translation of New Testament, 67; *Praise of Folly*, 274.
Ether, 106, 178.
Euclid (3rd C B.C.), 188n.9, 200; *Elements*, 188, 194, 195, 197, 234; Euclidian space, 111, 116, 118, 189.
Eudoxus (*ca.* 408–355 B.C.), 76.
Euripides, 48.
Eurythmy, 186.
*explicatio*, 96.

FABLE, 22, 257–58.
Fabricius, David (1564–1617), 206, 209.
Faith, 40–43, 63, 139.
Festugière, Andre, 194n.22, 279.
Feyerabend, Paul, 12 and n.23.
Ficino, Marcilio (1433–1499), 57, 60n.16, 112–14 and n.17, n.18, n.21; 133 and n.22, 144, 232, 246; *De lumine*, 130; *De sole*, 130 and n.10, 132, 136n.25; Latin translation of *Laws*, 56; *Theologica platonica*, 56 and n.4, 130 and n.11.
Filelfo (1398–1481), 103.
Fischer, Christopher (?–1511), 65n.33.
Fludd, Robert (1574–1637), 166–67, 246–52 and n.23, n.29, 36.
Foci, 208, 226, 284.
Fogliniani, Uberto (end of 15th C–1581), 233n.3.
Foix de Candale, Francois de (1502–1594), 144 and n.40, 153, 194 and n.20; commentary on the *Pimandros*, 136 and n.24, 144.
Formism, 18–19, 25, 26 105–06, 121, 124, 145–47, 152, 230, 282.
Foucault, Michel, 12, 24, 114n.21, 191.
Fracastoro, Girolamo (1483–1553), 53, 76n.8, 79, 120 and n.43; *Commentariolus*, 120 and n.43.

Francesco di Giorgio Martini (1439–1502), 136–37, 138.
Frey, Dagobert, 117n.37.
Friedrich, Hugo, 204n.6.
Froidmont, Libert (1587–1653), 156 and n.22, 160; *Vesta*, 156.
Frye, Northrop, 37, 46, 265n.19.
Funkenstein, Amos, 43n.25, 44n.28, 160n.30.

GAETANO DE THIEN (1387–1465), 47.
Gafurio, Franchino (1451–1522), *Practica musice*, 132, 133.
Galen (*ca.* 131–201), 96, 265.
Galilei, Galileo (1564–1642), 28, 46, 79–80, 130, 158, 198, 199–200, 204, 222–24 and n.47, 270, 273, 275; *Dialogue on the Great World Systems*, 280; *The Assayer*, 222–24.
Galilei, Vicenzio (1533–1591), 233 and n.4.
Garin, Eugenio, 60n.16, 63n.31, 129.
Gassendi, 248.
Gauricus, Pomponius (*ca.* 1480–1530), *De sculptura*, 94, 117 and n.35, 118.
*Geistesgeschichte*, 12.
Geminus (1st C B.C.), 38, 39n.12.
Gemma Frisius (1508–1555), 152–53.
Genette, Gérard, 29 and n.67, 30n.7, 60n.17, 61n.18, n.19, 62n.23, 228 and n.67.
Genre; *declamatio*, 50–52; dream, 253ff., 275.
Gentner, Dedre, 28, 146n.47.
Geocentrism, 14, 75ff., 91, 100, 102–03, 106, 112, 128, 131, 133–36, 145, 265.
Geometry, 91, 114–18, 135–36, 167ff., 170, 172–73 and n.16, 175ff., 186n.4, 194ff., 214, 222ff., 229–30, 233ff.
Gerl, Hanna Barbara, 70n.52, 71n.54.
Gerlach, W., 253n.1.
Ghiberti, Lorenzo (1378–1455), 69, 71–72.

Giese, Tiedemann (1480–1550), 50, 58.
Gilbert, William, (1540–1603), 152.
Gingerich, Owen, 43n.25, 63n.31, 91n.31 115n.26, 121n.45, 154n.10, 161n.32, 265n.19.
Godin, André, 63n.27.
Godwin, Francis, (1562–1633), *Man in the Moon* (1638), 254.
Godwin, Jocelyn, 250n.37.
Goulart, Simon (1543–1628), 95, 108n.2.
Gracian, Baltazar (1601–1658), *Criticón*, 274 and n.40.
Grant, Edward, 40n.15, 60n.20, 111n.8.
Grassi, Ernesto, 96n.51.
Gravity, 108ff.; *see also* Impetus.
El Greco, Domenikos Theotokopoulos (*ca.* 1540–1614), 210–11.
Greene, T., 61n.18, 62.
Greimas, A. J., 19–20.
Groupe *Mu*, 110n.4.
Guarini, Battista (1538–1612), *Pastor fido*, 255.
Guenther, L., 253n.1.
Guthrie, William, 113n.18.

HAASE, R., 217n.32, 232, 241n.14.
Hanson, Norwood R., 10, 92n.33.
Hardison, O. B., 265n.19.
Harmony, 73–74, 101, 111, 217–19, 232ff., 243ff., 250–52, 284; *see also* Polyphony.
Hauser, Arnold, 161n.33, 204n.3.
Heilbronn, John, 39n.12.
Heisenberg, Werner, 19.
Heliocentrism, 58, 61–62, 83–84, 90–91, 103, 123, 127ff., 152ff., 172, 219ff., 265 and n.19.
Hell, 132, 264.
Hempel, Carl, 8.
Heninger, S. K, 189n.14.
Heraclides Ponticus, 39.
Hermes Trismegistus, 127–28 and n.5,

130, 136n.24, 279.
*Hermetica*, 246.
Hermeticism, 57, 93n.36, 128 and n.5, 175.
Heron of Alexandria (2nd C–3rd C), 88.
Herrera, Juan de (1530–1597), *Discurso de la figura cubica* (1589), 181.
Hesse, Mary, 28.
Hipparchus (160–127 B.C.), 76.
Hippocrates (*ca.* 460–377 B.C.), 265.
Hippolytus, 64–65.
Hocke, Gustave, 172n.10, 196n.26, 203, 204n.6, 217, 246n.20.
Hofmann, Joseph E., 194n.20.
Holton, Gerald, 11 and n.17, 212n.25, 223.
Homocentrism, 76 and n.8, 120.
Hooykaas, Reijer, 65n.33.
Horsky, Z., 112.
Hübner, J., 178–79.
Human body, 57, 94, 95–96, 97ff., 188, 189; motion, 114–15, 210; representation, 57, 215; symmetry, 96, 156; *see also* Proportion, Renaissance art.
Humanism, 44, 63n.29.
Huyghens, Christian (1629–1695), 62n.24; *Codex Huyghens*, 114.
Hyperbola, 225ff.
Hypertextuality, 60ff. and n.17, n.18, n.23.
Hypsicles (2nd C B.C.), 194.

IDEOLOGY, 12.
Imitation, 61–62; *see also* Antiquity, *renovatio, similitudo.*
Impetus, 44 and n.29, 111, 144, 188, 209, 211; *see also* Gravity, Velocity.
Incommensurability, 43ff., 46, 137, 207–08.
Induction, 7n.1, 11, 45.
*Innovatio*, 60ff. and n.18, 193.
Intertextuality, 15, 24–27, 60n.17, 64, 94, 139, 187–89, 282–83.

Intervals, 186; harmonic, 234ff.; *see also* Proportion, Ratios.
Inversion, 271ff.
Irony, 21ff., 25, 29n.68, 35ff., 41–43, 46ff., 58–59, 152ff., 161; vertical, horizontal, 23, 35ff., 47.
Ivins, William, 214n.32.

Jacob, François, 18n.42.
Jakob, O., 60n.14.
Jamnitzer, Wenzel (1508–1585), 196, 200; *Perspectiva corporum regularum* (1563), 181, 196–97.
Jardine, Nicolas, 39n.12, 47n.39.
St. Jerome (*ca.* 340–420), 67.
Jesus, 141–44.
John, Gospel of (12:45, 14:6), 179.
St. John of the Cross (1542–1591), 140–41.
Juana Ines de la Cruz (1651–1695), 250n.36.

Kant, Emmanuel, 72.
Kappler, Claude, 100n.57.
Kempfi, Andrzej, 65n.33.
Kerszberg, Pierre, 111, 112n.14, 117n.33.
Kessels, A. H. M., 257n.8.
Kircher, Athanasius (1602–1680), 248n.26, 250–51 and n.36, n.37; *Ars magna lucis et umbrae, Misurgia universalis*, 250 and n.36.
Kirkwood, P. F., 253n.1.
Klaniczay, Tibor, 160n.31.
Klein, Robert, 117 and n.35, n.37, n.39; 138 and n.31, 160n.31.
Klibansky, Raymond, 57n.9, 111n.8, n.11.
Knowledge, 67ff., 288.
Koenigsberger, Dorothy, 232n.2.
Koyré, Alexander, 12, 24, 44n.27, n.29, 48n.42, 60n.14, 76n.6, 83n.18, 91, 111n.11, 114, 116n.31, 127 and n.2, 144n.42, 152, 161n.37, 163n.2, 204, 209n.17, 222 and

n.47, 241–42 and n.15, n.16, n.18.
Krajewski, Wladyslaw, 9.
Kris, Ernst, 94n.40.
Kristeller, Paul, 112n.14, 113n.20.
Kuhn, Thomas, 9, 57n.6, 76n.9, 79n.14, 156n.20, 281.
*Kunstkammer*, 286.
Kurz, Otto, 94n.40.

La Ceppède, Jean de (*ca.* 1548–1623), 255n.4.
Lacan, Jacques, 30.
Lactantius (*ca.* 260–*ca.* 317), 65–66, 93n.37.
Lafond, Jean, 274n.40.
Lakatos, Imre, 10n.15.
Landmann, Michael, 171n.9.
Latomus, Jacques Masson (1475–1544), 63.
Laudan, Larry, 10n.15.
Lausberg, H., 23n.50, 159n.28.
Le Molle, Roland, 138n.30.
Le Moyne, Pierre (1602–1671), 145.
Lear, J., 253n.1.
Lefèvre d'Etaples, Jacques (*ca.* 1455–1536), 63n.26, n.28.
Leibniz, Gottfried (1646–1716), "Tentamen anagogicum," 23.
Lenoble, Robert, 12, 19–20 and n.44, 24, 242n.18.
Leonardo da Vinci (1452–1519), 97, 101, 114–15 and n.25, nn.28–30, 117; *Libro del moto actionale*, 114.
Lerner, Michael, 39n.14, 44n.27, 50n.44, 85n.22, 158n.26.
Leurechon, Jean (1591–1670), 102.
Lindberg, D., 40n.15.
Line: *see* Curve and line.
Liszt, M., 253n.1.
Logical positivism, 8.
*logos*, 30.
Lomazzo, Giovanni Paolo (1538–1588), 94 and n.44, 97n.56, 118, 189, 210 and n.19.
Lotman, Juri, 26n.56.

Louis, A., 154n.12.
Lucian (*ca.* 125–192), *True History*,
    253.
Lukács, Georg, 47 and n.36.
Luminosity, 76, 79.

MACROBIUS, AMBROSIUS (end of 4th
    C), 256–57 and n.7; *Commentary on
    the Dream of Scipio*, 253, 256–58.
Macrocosm, 188–89.
Maestlin, Michael (1550–1631), 157
    and n.25, 158–59, 179; introduc-
    tion to *Narratio prima* (Rheticus),
    157, 185–86.
Magellan, Fernand de (*ca.* 1480–1521),
    205n.11.
Mahnke, Dietrich, 176n.22, 179n.25,
    188.
Maier, Annaliese 29, 31–33, 42n.24,
    43, 44nn.26, 45.
Mâle, Émile, 131n.14.
Mannerism, 117n.37, 160–62 and n.31,
    172–75, 182, 186–87, 188–89,
    196ff., 203ff. and n.6, 207–08,
    209–11, 217, 218, 219n.39, 226–28,
    247, 248, 250 and n.35, 255, 285.
Marcel, Raymond, 136n.24.
Marchal, P., 93n.36.
Marin, Louis, 228n.66, 275n.41,
    277n.43.
Maritain, Jacques, 255.
Marquetry, 197–99.
Maschler, C., 39n.12.
Mathematics, 38–39, 74, 90, 101, 111,
    133, 226, 249; *see also* Algebra,
    Astronomy.
Maula, E., 75n.5.
Mayr, Ernst, 18n.41.
Mazzoni, Jacopo (1548–1598), 262.
McGuire, J. E., 155n.18.
Méautis, Georges, 264n.18.
Mechanism, 17–18, 79, 152ff., 161.
Melanchthon, Philipp Schwarzerd
    (1497–1560), 47.
Melandri, E., *La Linea e li Circolo*, 7,

214n.32.
Menut, Albert, 41.
Menzel, Donald, 270n.35.
Mersenne, Marin (1588–1648), 129,
    224–27, 248; *First Book of Conso-
    nances*, 251; *Harmonie universelle*,
    251–52 and n.41; *Quaestiones in
    Genesim*, 128n.5, n.8, 224–26.
Mesnard, Jean, 200n.33.
Metaphor, 10, 27–30, 127ff., 230, 283.
Metonymy, 29, 80ff., 122, 230, 266,
    283.
Meyer, Bernard, 123n.46.
Meyer, Michel, 14n.10, 28, 31n.73, 29.
Miall, D. S., 146n.47.
Microcosm, 57–58, 74n.2, 94, 188.
Milton, John (1608–1674), 265n.19.
*mimesis*, 14, 16–19, 74.
Mittelstrass, Jürgen, 37n.4, 39n.12,
    206n.12, 212n.26, 212, 219n.39.
Model, 27–28, 208, 275ff.
Moesjgaard, K. P., 155n.19.
Molland, A. G., 125n.51.
Monochord, 247, 251.
Monstrosity, 47, 53–54, 64, 97–100,
    102, 146, 156–57, 160.
Montaigne, Michel de (1533–1592),
    48 and nn.41–43, 152n.4.
Moon, 76, 77, 79, 115n.29, 237n.12,
    261, 271n.37.
More, Thomas (1478–1535), 258;
    *Utopia*, 274.
Mossakowski, Stanislaw, 62n.25,
    132n.18.
Motion, circular, 79, 108–09, 111–12,
    121–25, 155, 205ff., 222; elliptical
    ("real"), 205ff., 218; lunar, 79, 154;
    natural, 107–09, 112, 179–80, 215;
    planetary, 61, 72, 74–80, 83–85,
    88, 89–91, 105, 118, 121–25, 128–
    29, 179, 205ff., 212ff., 237–38,
    285; solar, 42, 81; stellar, 158,
    285; sublunary, 106; supralunary,
    106, 215; terrestrial, 90, 106ff.,
    154; violent, 107–12.

Muecke, Douglas, 22n.49, 23, 35n.1.
Mulerus, Nicolas (1564-1630), 153.
Mysticism, 46n.35 and n.36, 140ff.,
    164-70, 190; see also Cabala,
    Numerology.
mythos, 13-14.

NAGEL, ERNEST, 17n.38, 18 and n.40.
Nancy, J.-L., 230n.79.
Narrative (in Kepler), 204-05, 259ff.
Negative mysticism, 47 and n.36.
Neoplatonism, 112-14, 124, 127n.2,
    250n.35, 282; see also Ficino.
Neugebauer, Otto, 76n.10, 125n.49.
Nicéron, Jean-François (1613-1646),
    102, 279.
Nicetas, 59.
Nicholas of Cusa (1401-1464), 60n.16,
    111 and n.9, n.11; 116, 175, 176
    and n.22, 179 and n.25, 196, 224.
Nicolas d'Autrecourt (14th C), 46.
Nicolson, Marjorie, 214n.29, 253n.2,
    n.3, 270 and n.35, 273n.39.
Nietzsche, Friedrich, 30.
Nifo, Agostino (1475-1538), 47.
Nominalism, 40, 43n.25, 46n.35, 47,
    71n.54.
North, John D., 200n.34.
Numerology, 165-66.

OBERMAN, HEIKO, 43n.25, 47n.38,
    50.
opifex, 54, 57.
Optics, 117; see also Perspective,
    Telescope.
Orbit, eccentric, 38-39, 74ff., 124-
    25, 239; elliptical, 203ff.; see also
    Deferent, Ellipse, Equant, Motion.
Oresme, Nicole (1323-1382), 39-43
    and n.28, 55, 61, 111 and n.9.
Organicism, 17, 25, 74, 83ff., 90-91,
    92, 95-101, 124, 152, 155-56, 159,
    185-86, 282.
d'Ors, Eugenio, 203, 204n.6, 208.
Osiander, Andreas (1498-1552), 49-52

and n.47, 79n.12, 152; Ad lectorem,
    50-52.
Oxymoron, 29, 215ff., 227-28, 279ff.,
    286.

PACIOLI, LUCA (ca. 1445-1514),
    188n.9, 189, 196, 200.
Palladio, Andrea (1508-1580), 233n.3.
Panofsky, Erwin, 46n.35, 69n.45,
    n.49, 70n.50, 94n.45, 96, 97n.55,
    n.56, 101n.62, 115n.24, 161,
    172n.10, 174n.18, 182, 204,
    211n.23, 222 and n.45, n.46,
    275n.42.
Pappus (3rd C), Mathematical Collec-
    tion, 177.
Parabola, 225ff.
Paracelsus, Theophrastus von
    Hohenheim (1493-1541), 151.
Paradigm, 9, 66, 67ff. and n.38, 194ff.
    236.
Paradox, 49ff.
Parallax, 85 and n.21, 156-57.
Paratextuality, 60n.17, 61, 62-63.
Pascal, Blaise (1623-1662), 214, 227-
    28 and n.65, n.66; Pensées, 228.
Patrizi, Francesco (1529-1597),
    112n.14.
Paz, Octavio, 250n.36.
Peirce, Charles, 7-8 and n.1.
Pelerin, Jean (1445-1524), 101.
Pepper, Stephen, 17-19 and n.35,
    89n.25.
Pereira, Benito (ca. 1460-1553), 47.
Periods, planetary, 75, 83-84, 88ff.
    and n.24, 91, 110, 118, 186, 217,
    223, 236ff., 240-41.
Perrault, Claude (1613-1688), 95.
perspectiva superior, 115-18, 129.
Perspective, 68-69, 101ff. 197ff.,
    244ff.
Perspectivism, 67-72, 114-18,
    207-08, 244-46.
Petrarch (1314-1374), De suis ipsius et
    omnium ignorantia, 57.

Peucer, Caspar (1525-1602), 154, 256 and n.7, 257, 265n.20.
Peurbach, Georg von (1423-1461), 79n.13; *Theoricae novae planetarum*, 47 and n.40.
*phantasm*, 256-57.
Phillips, J. P., 200n.33.
Philo (20 B.C.-A.D. 54), 171.
Philology, 63-67 and n.34, 69-70, 165.
Physics, 110-11, 145ff., 163; Aristotelian, 38, 53, 106, 109-10, 140, 155.
Piaget, Jean, 30-31.
Pico della Mirandola, Jean (1463-1494), 47n.39, 57, 130, 144; *Heptaplus*, 130n.11, 144.
Piero della Francesca (1416-1492), 101, 196; *De quinque corporibus* (1492), 181.
*Pimandros*, 128n.5, 136, 144 and n.40.
Planets, "inner" and "outer," 77, 111, 121, 146; volume of, 241-42. *See also* Motion, Space, Velocity.
Plato (428-347 B.C.), 47n.39, 127n.2, 128, 171, 175, 184, 194, 227, 232; *Cratylus*, 68, 69; *Laws*, 55-56, 111n.12; *Phaedrus*, 36; *Republic*, 196.24; *Symposium*, 270n.34; *Timaeus*, 35-36, 55, 111-12 and n.12, 187 and n.5, 188, 189, 197, 242n.19.
Platonism, 37n.4, 47, 56, 93, 115n.27, 129-30, 138, 171ff., 189, 196-97, 222 and n.47, 246, 282.
Plattard, Jean, 66n.36, 153n.7.
Play, 22, 41ff.; God's, 163; in Kepler, 164-70, 190, 257ff.; in Oresme, 41-43.
Pliny the Elder (23-79), 128.
Plotinus (*ca.* 203-270), 171; *Enneads*, 175.
Plutarch (*ca.* 48-*ca.* 122), 59-60, 63n.30; *De facie in orbe lunae*, 253, 264.
Pocock, John, 63 and n.29.

Poetics, 14ff., 20.
Poinset, L., 200
Poliziano, 56n.4, 63; *Miscellaneorum centuri secunda*, 64 and n.32.
Polygons, 181, 188ff., 234-36, 247; hierarchy, 188; nesting, 185, 195-96, 199-200, 241, 243; primary, secondary, 194.
Polyhedrons, 181, 184-85, 189ff., 193ff., 241, 284.
Polykleitos (5th C B.C.), 96; *Canon*, 96.
Polyphony, 218-19, 239-40.
Pontano, Giovanni Giovano (1429-1503), 47.
Pope, 62, 73; Leo X, 67; Paul III, 67, 73.
Popkin, Richard, 47n.37.
Popper, Karl, 8.
*potentia (absoluta, ordinata)*, 43-46 and n.25, n.33.
Poulet, Georges, 211n.24, 255.
Praetorius, Johannes (1537-1616), 50 and n.44, 51, 52 and n.52, 153-54, 198-99.
Prague, 204, 231, 265, 266.
Price, Derek, 61n.21, 92n.32, 125n.50.
Proclus (412-485), 120n.42, 175, 194 and n.22; *Commentary on the First Book of Euclid*, 175-76, 182.
Proportion, 93, 94ff. and nn.42-44, 155ff., 185-86, 187-88, 207-08, 209, 214; mathematical, 101-02, 240ff.; of matter, 241-42.
*propter nos*, 55-59, 72, 157, 158-60, 174, 243-46, 282.
Ptolemy, Claudius, (2nd C), 35, 36-38, 40, 55, 56, 60ff. and n.21, 76 and n.6, 77-79, 83-84, 91-92, 96, 103, 118, 120, 122, 123, 137, 184n.2, 190n.16, 206, 215, 220, 281; *Almagest*, 36-38, 56, 60, 61-62, 83, 85, 120; *Harmonies*, 231, 256n.7.

Pyrrho (*ca.* 360-270 B.C.), 48, 49n.43.
Pythagoras (*ca.* 580-*ca.* 500 B.C.), 184, 219, 232-33.
Pythagoreanism, 62 and n.25, 93, 96n.51, 136n.23, 138, 166, 175, 187, 189, 191-93, 196-97, 200, 264n.18, 267.

QUINTILIAN, MARCUS FABRICIUS (*ca.* 30-95), 117.

RABELAIS, FRANÇOIS (*ca.* 1484-1553), 111n.9, 278.
Racine, Jean (1639-1699), 14.
Raimondi, Ezio, 224n.56, 228.
Ramnoux, Clémence, 145.
Ramus, Pierre de la (1515-1572), 50n.44.
Randall, John, 43, 44n.26.
Raphael, Raffaelo Sanzio (1483-1520), 115 and n.26, 138n.28; *Disputa della Sacramento* (1509), 132-33.
Ratios, 237-38, 240-42.
Ravetz, Jerome, 61n.21.
Raymond, Marcel, 217n.36, 262n.15.
Redondi, Pietro, 224n.53.
Redondo, Augustin, 274n.40.
Regiomontanus, Johannes Müller (1436-1476), 79n.13; *De triangulis omnimodis*, 188.
Reichenberger, K., 108n.2.
Reinhold, Erasmus (1511-1553), 59n.12, 121, 154; *Prutenic Tables*, 121.
Reiss, Edmund, 47n.36.
Reiss, Timothy, 67-68, 280n.47.
Renaissance art, 25, 68-70, 93ff., 101, 114-18, 171-72, 174-75, 187, 196, 226-27.
*renovatio*, 59ff. and n.18, 62ff., 123, 151-52, 202.
Representation, 12, 58, 68ff., 101ff., 120ff., 159-62, 167ff., 179-80, 225, 281; astronomical, 37ff., 90ff., 159-62, 207ff., 219, 221; mathe-matical, 56, 101-02, 223; philo-sophical, 227-28.
Resemblance, 112-13, 167, 173, 189-91, 230, 261.
Rheticus, Georg Joachim (1514-1576), 49, 50n.44, 58, 75, 79, 88, 91, 121, 157, 185; *Narratio prima*, 13, 40n.16, 50n.45, 58n.11, 60, 62, 152, 185.
Riccioli, Giambattista (1598-1671), 160, 251; *Almagestum novum*, 145, 156-57.
Richter, J.-P., 115n.30.
Ricoeur, Paul, 13, 14 and n.30, 28n.60.
Ridgely, Beverly, 153n.7.
Riffaterre, Michael, 26n.57, 129n.8, 236n.9.
Righetti, M., 131n.7.
Rose, Paul, 44n.27, 63n.31.
Rosen, Edward, 41, 42, 50n.44, 59n.12, 62n.25, 93n.36, 127n.1, 128 and n.5, 157n.24, 253n.1, n.2, 264.
Rosselli, Annibal (*ca.* 1524-1600), 136n.24.
Rotation, solar, 155; terrestrial, 40ff.
Rothmann, 94n.38, 155.
Rubka, E., 130n.10.
Rudolf II, 204, 205n.10.
Russell, J. L., 223.
Ruysbroeck, Jean de (1293-1381), 46.

SACHS, E., 189n.14.
San Carlo alle Quatro Fontane, 226.
Sant' Andreo al Quirinale, 226.
Santillana, Giorgio de, 116.
Sarduy, Severo, 26-27, 115n.26, 203, 208, 227n.64, 275n.42.
Scaliger, Julius Caesar (1484-1558), 172.
Schavernach, H., 232n.2, 248n.25.
Schmitt, Charles, 47n.37.
Scholasticism, 35, 43, 57, 61, 63, 66, 160.

Scholder, Klaus, 152n.3.
Schön, Erhard (*ca.* 1491–*ca.* 1542), 188, 196.
Schwarzenberger, G. von, 204n.8.
Seck, Friedrich, 184n.2.
Segonds, A., 50n.44.
Segré, C., 26n.56.
*semiosis*, 16, 19ff., 67–72, 162, 163ff., 177ff.
Serres, Michel, 205n.10, 227n.65.
Sextus Empiricus (3rd C), 47, 48.
Seznec, Jean, 132n.18.
Shakespeare, William, 28.
Shapere, Dudley, 10.
Shearman, John, 210n. 20, n.21, 254n.4.
Signs, linguistic, 68–69, 164–70; pictorial, 68–69; *see also* Representation, *semiosis*.
*similitudo*, 61, 74.
Simon, Gérard, 13, 73–74, 181, 205n.11, 234n.7, 284–85.
Simplicius (1st half of 6th C), 39n.12.
*simulacrum*, 69n.46, 172 and n.15.
Skepticism, 47–49, 70, 182n.32; *see also* Irony.
Snecanus, 47 and n.7.
Sneed, J., 9.
Somatopoetic, 180–81; *see also* Cosmopoetic.
*somnium*, 256–57.
Sophocles (*ca.* 496–406 B.C.) 127 and n.1, 128.
Space, 110–11, 113, 115–18, 154ff., 178, 183, 187, 196–99; proportion, 155ff.; sublunary, 110–13, 131, 140–41, 147; supralunary, 110–13, 131, 147, 284; *see also* Center, Euclid, Polyhedra.
*species immateriata*, 209–11, 284–85.
Sphere(s), 61, 85ff., 105ff., 112–14, 146, 177ff., 183, 247.
Spinoza, Benedict (1632–1677), 251n.40.
Spitzer, Leo, 181n.29.

Stahlman, William, 62n.25.
Stars, 85ff., 92, 93, 158, 179, 220–21; fixed sphere, 85–88, 92–93, 156–57, 158–59, 178, 179, 211, 220, 285; size and distance, 158–59; *see also* Parallax, Space.
Steck, M., 182.
Stengers, Jean, 152n.3, 159n.29.
Stevin, Simon (1548–1620), *De Hemelloop*, 219–21.
Stimson, Dorothy, 152n.1.
Stoer, Lorenz (*fl.* 1550–1567), 196; *Geometria et perspectiva* (1567), 181.
Stoics, 57.
Sufficient reason, 163, 170–71.
Sun, 14, 75–76, 77, 81, 88, 90–91, 117–18, 127ff., 140ff., 178, 179, 220–21; as source of motion, 128, 211, 220; centrality, 88, 117–18, 127, 130, 145ff., 220–21, 283; immobile, 39, 143ff., 160–61, 179, 211; symbolism, 127ff.; *see also* Impetus, Heliocentrism.
Suvin, Darko, 254n.3, 271n.36.
Swerdlow, N., 125n.49.
Symmetry, 20–21, 73–74, 77, 80, 88–89, 90, 92ff. and n.56, 105, 114–15, 117–18, 128, 146, 153, 155ff., 185–86, 217, 283, 285; terminology, 94–95.
Synecdoche, 29, 106, 110–11, 114–15, 122–23, 146, 283.
Synesthesia, 250n.38.
Sypher, Wylie, 204n.3.

TASSO, TORQUATO TASSOLE (1544–1595), 198, 222.
Telescope, 68, 158, 199–200.
*telos*, 16, 21, 35.
Terminology, of dreams, 256–57; of symmetry, 94–95.
Tesauro, Emanuele (1592–1675), 224, 227 and n.61, n.62, 228–29, 251–52.
Text, 36, 202, 257ff.; *see also* Hyper-,

Hypo-, Intertextuality.
Theocentrism, 137–39, 144.
Theology, 39, 43ff., 65–67.
Theory, 9ff., 152–53.
Theresa of Avila (1515–1582), 140n.36; *Moradas*, 140.
Thomas Aquinas (1225–1274), 40 and n.16, 42n.23, 43n.25, 47, 171–72.
Thorndike, Lynn, 151n.1, 153n.7.
Toulmin, Stephen, 9, 204n.3.
Tournon, A., 49n.43, 51n.49, 52 and n.51.
Trinity, 133, 139, 163, 178–79, 211–12, 214, 230, 247, 251, 284.
Tropology, 15, 21–22, 27–31, 80–81, 105, 123, 250n.36, 265, 283.
Tuzet, H., 202.
de Tyard, Pontus (1521–1605), 76n.7, 79n.12.

Uccello, Paolo di Dono (*ca.* 1396–1475), 197.
Uspenskij, B., 26n.56.
Utopianism, 138, 254, 271, 274–78.

Valla, Lorenzo (1407–1457), 57 and n.8, 63, 70–72 and n. 51, n.52, n.54, 165; *Adnotationes*, 165n.33.
Vasari, Giorgio (1457–1511), 197.
Velocity, 74–75, 121–123, 147, 158 and n.26, 218, 237.
Vergil, *Aeneid*, 80–81.
Vernant, Jean-Pierre, 12.
Vertical hierarchy, 131ff., 137, 145f., 282.
Vico, Giambattista, 30; *New Science*, 21–22.
Vigenere, Blaise de (1523–1596), 94–95.
Villard de Honnecourt (13th C), 96 and n.53.
Vitruvius, Marcus Pollio (1st C b.c.), 88n.23, 94, 95, 186 and n.4, 188 and n.9.
Vives, Juan Ludovico (1492–1540),

71–72, 165.

Waddington, Raymond, 130n.11, 144n.41.
Wagner., J.-M., 172n.13, 174n.14, 255n.5.
Walker, D. P., 199n.31, 233n.4, 235n.8.
Waswo, Richard, 70n.52, 71n.55.
Waterbolk, E. H., 153n.5.
Wellek, Rene, 250n.38.
Westman, Robert, 50n.47, 129n.8, 153n.6, n.9, 154n.10, 155n.18, n.19, 157n.25, 187n.5, 188n.8.
White, John, 69n.48, 101n.63.
Wilkins, John (1614–1672), World of the Moon (1638), 254.
Wilson, Curtius, 208–09.
Wind, Edgar, 132n.18.
Wirth, Jean, 139n.32.
"Wittenberg interpretation," 121f., 154 and n.10.
Wittkower, Rudolf, 138, 187n.6, 233n.3.
Wolfflin, Heinrich, 204n.7.
Wrightsman, B., 50.
Wunderkammer, 286.

Xenophanes, 48.

Yates, Frances, 127–28 and n.4, 167n.5, 246.

Zarlino, Giuseppe (1517–1590), 233 and n.3.
Zeno, 48.
Zilsel, Edgar, 113n.18.
Zinner, Ernst, 50n.44, 152n.1.
Zodiac, 76, 77, 83, 84, 164 and n.3, 190.
Zohar, Elie, 10n.15.
Zorzi, Francesco Giorgio (1460–1540), 246, 247; *De harmonia mundi*, 179.
Zuccari, Frederico (1540–1609), 172 and n.10; *Idea*, 191.